Kari Stenman

Colour profiles:
Andrzej M. Olejniczak

Finnish Trainer Colours
1930–1945

Published in Poland in 2023
by STRATUS s.j.
Żeromskiego 6A
27-600 Sandomierz, Poland
as
MMPBooks,
e-mail: office@mmpbooks.biz
© 2023 MMPBooks.
http://www.mmpbooks.biz

ISBN
978-83-67227-09-4

Editor in chief
Roger Wallsgrove

Editorial Team
Bartłomiej Belcarz
Robert Pęczkowski

Text and research
Kari Stenman

Colour profiles
Andrzej M. Olejniczak
Karolina Hołda

Proofreading
Roger Wallsgrove

DTP
Bartłomiej Belcarz

Printed by
Wydawnictwo Diecezjalne
i Drukarnia
w Sandomierzu
www.wds.pl

PRINTED IN POLAND

Table of Contents

Foreword 3
Aviation Schools 5
Basic Trainers 25
    Letov Š-218A *Smolik* 26
    VL *Sääski* I, II, IIA and IV 40
    De Havilland D.H. 60X Moth and D.H.82 Tiger Moth 48
    VL *Viima* I, II and IIB 60
    Focke-Wulf Fw 44 J *Stieglitz* 78
Advanced Trainers 91
    Gloster Gamecock I and II 92
    ASJA *Jaktfalk* II 100
    Bristol Bulldog II and IVA 104
    Gloster Gauntlet II 114
    Polikarpov I-15bis 126
    VL *Pyry* I 130
    Fokker D.XXI 146
    Fokker C.X 158
    FIAT G.50 160
Twin-Engine Trainers 167
    Avro Anson I 168
    Ilyushin DB-3M 174
    Hanriot H.232 178
    Bristol Blenheim I 182
    Airspeed AS.6E Envoy 190
    Tupolev SB 192
Gunnery Trainers 199
    VL *Tuisku* 200
    Aero A-32GR 212
    Blackburn Ripon IIF 220
    Fokker C.V 228
    Koolhoven F.K.52 242
Liaison Aircraft 245
    Beechcraft C-17L Traveler 246
    Cessna C-37 Airmaster 248
    Desoutter II 250
    Fairchild 24G 252
    Junkers A 50 Junior 254
    VL *Kotka* 255
    Fieseler Fi 156 K-1 *Storch* 258
    Polikarpov U-2 262
Camouflage & Markings 266
Abbreviations 272

Acknowledgements

The documents consulted came from the Finnish National Archives and its War Archives branch. These were the individual aircraft files, State Aircraft Factory repair reports, training unit war diaries and various reports. Photographs were obtained from the Finnish Air Force, Finnish Air Force Museum, Defence Force, National Archives, War Museum and Finnish Aviation Museum collections. In addition to the author's collection photos have been obtained from the fellow researchers Lassi Eskola, Pentti Manninen, Klaus Niska, Kyösti Partonen, Eino Ritaranta and Veikko Salo.

Also the following veterans or their families gave access to their logbooks and photos: L. Bergman, A. Bremer, K.L. Bremer, P. Ervi, R. Hyvärinen, E. Jaakkola, A. Juurinen, A. Kaila, E. Laiho, R. Lampelto, E. Laukkanen, A. Nieminen, A. Nissinen, U. Nurmi, P. Nurminen, L. Piippo, M. Piisilä, O. Riekki, R. Rosenberg, K. Tuomikoski, L. Volanen and T. Vuorinen.

A special mention is due to my fellow historian Kyösti Partonen and his input to cover the early camouflage schemes.

The last but not least thank you goes to the illustrators Andrzej M. Olejniczak and Karolina Hołda, whose artwork are simply the best available on the Finnish topics.

Foreword

Trainers for a New Air Force

Finland declared independence from Czarist Russia on 6 December 1917. Soon after that a civil war began between the white and red participants, ending in May 1918 in a victory for the whites, with some support from German troops coming from Estonia. All Russian garrisons and air bases were disarmed and the aeroplanes confiscated. While the Great War still lasted, several aircraft were bought from the Germans in Tallinn, Estonia. A good number of planes also defected from Russia.

When the central powers lost the Great War, Germany as a source of aircraft came to an end. Now France became a significant supplier, from whom a good number of Brequet 14s and Caudron G.3s were bought in 1919–1920. During the first five years of independence the total number of aircraft was not substantial and practically every one of the over thirty different types was used in training.

Hansa floatplane coded IL-118 of MaalK at Santa-hamina on 2 October 1929. The Black chevron dates back to IlmK and a tripod serves as a plane marking. (Finnish Air Force)

The first actual trainer purchase consisted of 30 Caudron C.60s, which were delivered in 1922–23, in addition to a licence to build them at the new Aviation Force Aircraft Factory at Suomenlinna, outside Helsinki. Licence production was carried out in 1927–28 and 34 aircraft were produced.

The first dedicated training unit, *Ilmailukoulu* (Aviation School) was established at Santahamina near Helsinki on 1 March 1921. Santahamina's location by the sea was ideal for seaplanes and the main type in this duty was the Hansa (licence-built Hansa Brandenburg W.33), which entered service in 1923, ending six years later.

From 1923 onwards the main land plane trainer type was the Caudron C.60 until the Czech Letov *Smoliks* started to replace them from April 1931. But the Caudrons still served until August 1936.

Santahamina's location on an island near the capital Helsinki limited the development and growth of the flying school. New purpose built facilities were situated in central western Finland at Kauhava, which was declared operational on 6 July 1929. The flying training was now totally land based.

Before the school's transfer to Kauhava, a total of 158 men had received the military pilot's licence. When the Continuation War ended in September 1944, a total of 1,068 men had gone through the training at Kauhava and became a licenced military pilot. On top of this came the observers, wireless operators and gunners.

Aviation Schools

Lentosotakoulu

Primary flying training was transferred in 1929 to Kauhava in central western Finland, when the *Ilmailukoulu* (IlmK, Aviation School) moved there to purpose-built premises. On 1 January 1938 with the Air Force re-organization the name of the Aviation School was changed to *Ilmasotakoulu* (ISK, Air Fighting School). LtCol Toivo Somerto led the school until the mobilization for the Winter War when he was posted to command *Lentorykmentti* 4 (Aviation Regiment 4) on 7 October 1939. The previous commander of *Lentorykmentti* 2, LtCol Rainer Ahonius, was appointed to lead the ISK. The number of personnel at this point stood at 550.

The main type used in elementary training was the *Smolik*, also accompanied in 1939 by *Viimas* built by the State Aircraft Factory. Observer and gunnery training was carried out on *Tuiskus* in the first place and also on *Sääskis* and Aero Jupiters. Just before the Winter War the school handed over a mixture of aircraft to supplementary regiments, receiving appropriated civil aircraft instead. The aircraft strengths of the school can be seen in the table within the appendix.

Most primary flying training of the Finnish Air Force took place at Kauhava in central western Finland, where Lentosotakoulu (Aviation Fighting School) was located. This name was adopted on 1 March 1941. The aerial photo of the base was taken on 20 August 1937. The airfield was square shaped and measured 900x1000 metres. The lighter square area was for take-offs and landings. (Finnish Air Force)

Smoliks are pushed out to the airfield at Kauhava on 13 August 1936. It was done by man-power as one Smolik weighed 628 kilos. At this point the school was named Ilmailukoulu (Aviation School) but on 1 January 1938 it became Ilmasotakoulu (Air Fighting School). (Finnish Air Force)

Training accidents were not uncommon. Here Smolik SM-132 ran into Sääski SÄ-136 during a take-off from Kauhava on 19 May 1937. Both aircraft were repaired. (Finnish Air Force)

Smolik SM-157 of ISK in a field overhaul at Joroinen in summer 1939, when the Finnish Air Defence Association held pilot courses for civilians. Ten Smoliks of ISK were used. (Author's collection)

Winter War

At the outbreak of the Winter War on 30 November 1939, the Air Fighting School operated from Kauhava with Mänkijärvi and Laajalahti as additional bases. The training programme was continued but with a tighter schedule. It was soon discovered that this made no contribution. Early in the conflict the personnel fully occupied the front line squadrons and since the losses were considerably lower than calculated, there was no need for new airmen. Therefore the supplementary units were full and could not take more students to be trained for front-line duties. Thus graduated airmen remained in various posts at the Air Fighting School.

At the beginning of the mobilization on 6 October 1939 the school had 107 students. The pilot training consisted at this point of about 80 flying hours. During the Winter War the school flew almost 10,000 hours and produced the following airmen:

Cadet Course 22	17 pilots	
Reserve Officer Course 9	36 pilots	10 observers
NCO Pilot Course 9	13 pilots	
Reserve NCO Pilot Course 6	10 pilots	

A Smolik of ISK is refuelled by hand-pump from 200 litre drums at Joroinen in summer 1939. The engine is a 145 hp Walter Mars radial. (Author's collection)

During the Winter War the Air Fighting School arranged Observer Courses 1–3, of which the last ended on 20 May 1940. Eighteen active and 64 reserve officers were trained. During the intermediary peace Observer Course 4 was concluded on 8 June 1940, completed by 32 reserve officers.

After the outbreak of the hostilities separate war pilot courses were established. The preference was for students with elementary flying training or military service. The first of these four courses began on 5 December 1939 and the last ended on 20 July 1940. Pilots graduated from these courses:

Pilot Course 1	84 pilots
Pilot Course 2	63 pilots
Pilot Course 3	63 pilots
Pilot Course 4	61 pilots

The experiences in the Winter War showed that the training had been correct, the emphasis in developing good shooting skills had brought results. In July 1940 a renewed pilot training programme was began at *ISK* with more emphasis on shooting skills and tactical manoeuvring and added flying practises. The training period for both reservist officers and non-commissioned officers was to be 18 months, which contained about 140 flying hours.

The name of *Ilmasotakoulu* was changed to *Lentosotakoulu* (*LeSK*, Aviation Fighting School) on 3 January 1941, which better represented its task, since the school programme did not contain anti-aircraft or air surveillance duties.

Two generations of pilots. Left the student Olli Kepsu and the instructor, his father WO Taneli Kepsu, in front of Smolik SM-128 at Kauhava on 20 April 1940. (Finnish Air Force)

Another mishap for LeSK at Kauhava on 18 January 1941. Smolik SM-153 landed on Stieglitz SZ-7, with no casualties. Again, both aircraft were repaired. (Finnish Air Force)

Ready for a training flight at Kauhava in a Stieglitz during spring 1941. Only the instructor's rear cockpit has instruments. (Finnish Air Force)

The aircraft assortment of LeSK on exhibit at Kauhava in July 1943. From left, Gauntlet, Jaktfalk and two Pyrys. Further back are a DB-3M and Tuiskus. The building at left is the school headquarters, next is hangar No. 2, opposite hangar No 1. (SA-kuva)

Continuation War

At the outbreak of the Continuation War on 25 June 1941 the school continued to operate at Kauhava. The trainees were split and various courses operated from Mänkijärvi and Laajalahti bases plus, in winter, Lappajärvi ice base. If necessary Lestijärvi, Ylivieska, Mustasaari, Siikakangas and Vesivehmaa airfields under the school's control could also be used.

During the Continuation War the school flew just over 56.000 hours and primary training was given to 351 pilots, split as follows:

Officer Course 10	29 pilots
Officer Course 11	25 pilots
Officer Course 12	32 pilots
Officer Course 13	85 pilots
NCO Pilot Course 10	20 pilots
NCO Pilot Course 11	31 pilots
NCO Pilot Course 12	30 pilots
Pilot Course 5	52 pilots
Pilot Course 6	47 pilots

Additionally the 56 students of Pilot Course 7 managed to get the elementary flying training. All others except Pilot Course 6 students also underwent the supplementary flying training.

The Aviation Fighting School also trained instructors and from five courses 47 graduated during the Continuation War. No observer courses had been arranged since the Winter War, but on 20 July 1942 they were initiated. 148 officers graduated these 3–4 month courses during the war:

Observer Course 5	24 observers
Observer Course 6	5 observers
Observer Course 7	30 observers
Observer Course 8	32 observers
Observer Course 9	30 observers
Observer Course 10	27 observers

The hostilities ended on 4 September 1944 in a cease-fire and the Moscow truce a fortnight later.

A fellow pupil is fixing the helmet of his colleague before a flight in Smolik SM-151 of LeSK at Kauhava in July 1943. Fuel spillage from the upper wing tanks has corroded a lot of the orange paint on the wings. (SA-kuva)

One gunnery student has finished his shooting and exits a Tuisku of LeSK at Kauhava on 29 July 1943. The second student waits for his turn, carrying an ammunition drum. (SA-kuva)

After the War

According to the truce terms the Finnish military forces were to be demobilized by 4 December 1944. The reservists began leaving on 7 November 1944 and the demobilization was executed by the deadline. At this point the school had 120 elementary or advanced trainers.

The air force HQ cut down the amount on 23 January 1945 to 70, which consisted of 20 *Stieglitzes*, 15 *Pyrys*, 10 FIAT G.50s, 10 Fokker D.XXIs, 10 *Viimas* and 5 *Tuiskus*.

Flying could not commence before 27 July 1945 since the Allied Supervision Commission, which arrived just after the truce and headed by the Russians, banned all flying except the missions against the Germans or transfers to peacetime bases. The same commission put pressure on the Finnish government to change the blue swastika insignia to a cockade, which was carried out on all aircraft by 1 April 1945.

Supplement Regiments

According to the plans drawn up in 1935 and the subsequent updates, in case of a mobilization two advanced training squadrons were to be established at every air station. After the cessation of possible hostilities these training formations were to be disbanded in due course. When the Air Force was reorganized from permanent air stations into mobile regiments on 1 January 1938, two training squadrons were to be set up in every regiment. Their primary task was to train the airmen coming from the reserves to meet the requirements of the front line squadrons.

In the mobilization for the Winter War, or the General Rehearsals, one supplementary regiment was established on 9 October 1939 in every three existing aviation regiment and a training squadron to the Detached Squadron subjected to maritime duties. The personnel and equipment came mainly from the mother regiment. The necessary additional aircraft were drawn from the Air Fighting School and appropriation of civil aircraft. The lack of resources allowed only one training squadron to be formed in each regiment.

Counting hits in the target sleeve at Laajalahti camp in summer 1943. Behind is Tuisku TU-166 of LeSK, which performed the gunnery practice. (Author's collection)

T-LentoR 1

Lt.Col Lars Schalin was made commander of *Täydennyslentorykmentti* 1 (Supplement Aviation Regiment 1) and Capt Urho Toivonen the leader of the training squadron. The base was set up at Karvia and three Aero Jupiters, five *Sääskis*, two *Kotkas* plus one Fokker C.X and one *Tuisku* comprised the equipment. On 17 January 1940 Maj Arvo Nisonen was put in command of both *T-LentoR* 1 and the training squadron. In mid-February 1940 the squadron received two Fokker C.Vs, plus five more by the end of the Winter War.

By an order from the air defence on 11 May 1940 *T-LentoR* 1 was cut down to a single squadron and also named *T-LLv* 17 (Supplementary Squadron 17) commanded by Maj Nisonen. On 26 May 1940 Capt Armas Viherto took over the command and on 20 July 1940 the unit was to be disbanded by 15 August 1940. The reservists were demobilized and rest of the personnel sent to Aviation Regiment 1. The aircraft were handed over to other units.

*T-Lento*R 1 received 164 students for advanced training, 40 thereof being transferred in January 1940 to other supplementary regiments. During the Winter war it trained 49 pilots, 12 observers an 6 gunners for front-line squadrons. Before the disbandment another 57 airmen were trained.

T-LentoR 2

Maj Einar Nuotio was placed in command of *Täydennyslentorykmentti* 2. The training squadron was also named in this connection as *T-LLv* 29 with Capt Kaarlo Lejon in command. The unit was based at Parola and it was equipped with five Gamecocks, four *Tuiskus* plus one *Kotka* and a Moth. More aircraft were received twice, on 15 December 1939 three *Jaktfalkens* and two Bulldogs arrived from Sweden and another six Bulldogs from *LLv* 26 on 2 February 1940. Due to a large number of students (over 60) half of the unit moved in late January 1940 to Tyrväntö.

T-LentoR 2 was disbanded on 29 March 1940 by the air force C-in-C order and its headquarters formed the HQ of the new *LeR* 3. *T-LLv* 29 was disbanded at the same time and the whole was placed in Supplementary Squadron 35 (*T-LLv 35*) under *LeR* 3. Capt Eino Carlsson was appointed in command. The base was set at Parola.

T-LentoR 2 received for training 281 students, of which 39 were foreigners. During the Winter war 132 pilots graduated to front-line units, 26 of these being foreigners. The remaining 144 students were trained during the intermediary peace.

Ground crew manually pumping fuel into Tuisku TU-167 *of LeSK at Laajalahti in summer 1943. The fuselage tank had the capacity of 225 litres, which gave an endurance of 6½ hours. (Author's collection)*

T-LLv 35, formed for advanced training, was disbanded on 15 July 1940 and its personnel and equipment were transferred to *Lentolaivue* 34, also part of *LeR* 3. On 27 July 1940 Capt Olavi Ehrnrooth was appointed in command. The squadron was based at Paola until 18 June 1941, when it was subjected to the re-established *T-LLv* 35, now with Major Ehrnrooth as the commander.

Bulldog BU-66 and personnel of T-LLv 29 at Tyrväntö in March 1940. This unit was an advanced training squadron under T-Lento R 2 and gave future fighter pilots supplementary training. (Author's collection)

Plane inspection of LeSK held at Kauhava on 8 May 1944. Smoliks are in the first line, Tuiskus and Viimas in the second and more Tuiskus in the third. The nearest machine is TU-161. (Finnish Air Force)

Aero A-32 serialled AEj-57 of LeSK at Kauhava in early January 1940. The camouflage of Olive Green upper and aluminium dope undersides was specified for trainers on 12 October 1939 and is very fresh here. (Author's collection)

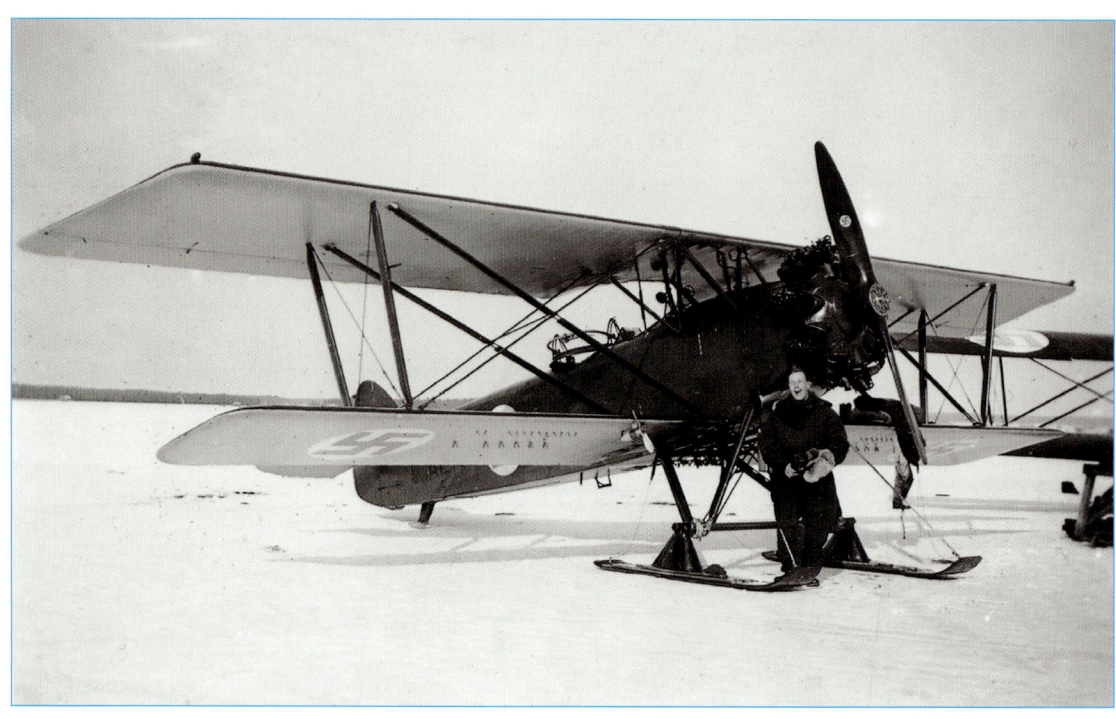

T-LentoR 4

Capt Armas Eskola was appointed to command *Täydennyslentorykmentti* 4 and its training squadron. The base was set at Luonetjärvi and three Ansons plus one *Kotka* and one *Tuisku* served as the aircraft. Two Dragon Rapides were also appropriated from Aero Oy.

On 16 January 1940 Eskola was replaced by 1Lt Tauno Vasamies. On 29 March 1940 the training squadron became *T-LLv* 47 and it was disbanded on 15 July 1940. The personnel and aircraft were transferred to other LeR 4 squadrons.

T-LentoR 4 received 160 students, of which 17 pilots, 30 observers and 17 gunners graduated during the Winter War. Another 92 students graduated during the intermediary peace time.

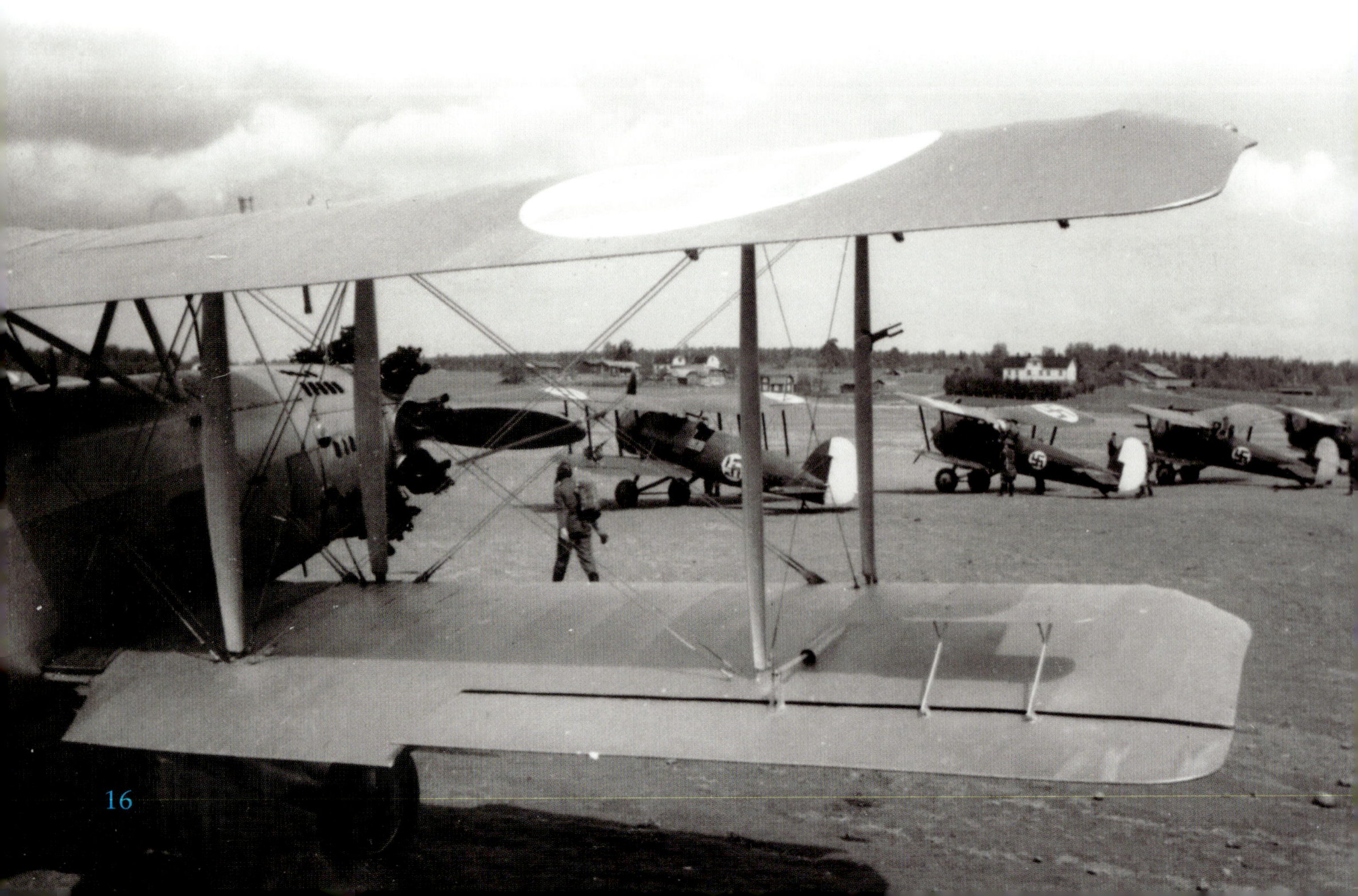

Gauntlets and Bulldogs of advanced training squadron T-LLv 35 at Parola in June 1940. The rudders are white, indicating trainers. The nearest machine is the unit's hack Kotka KA-145. (Author's collection)

A solitary Fokker C.X seri-alled FK-79 shortly before it was assigned to T-LLv 25 on 9 July 1941. This was a temporary advanced training squadron, which was merged into T-LLv 35 on 1 October 1941. (Finnish Air Force)

T-LLv 39

Capt Lennart Collin was posted to command *Täydennyslentolaivue* 39. The 1st Flight participated in combat missions with two Coast Guard Junkers F 13 planes based at Aaland. The 2nd Flight trained at Turku and from 31 December 1939 at Pori, with two *Tuiskus*, two Ripons and one Moth. In addition three appropriated civil aircraft came to the inventory – Waco, Cessna and Fairchild.

By the Air Force commander's order of 29 March 1940 *T-LLv* 39 became *T-LLv* 9 and Capt Lauri Bremer was appointed in command. The squadron was disbanded on 20 July 1940 and the remaining parts were transferred to the Detached Squadron (*ErLLv*).

T-LLv 39 trained during the Winter War 15 pilots and 8 observers from 44 students. During the peace another 21 airmen were trained.

Leaders of T-LLv 35 *at Parola on 10 July 1941. From left 1st Flight leader Capt Pekka Käär, squadron commander Maj Olavi Ehrnrooth, war correspondent von Wille-brandt and 3rd Flight leader 1Lt Jorma Visapää. Behind is Pyry PY-1. (SA-kuva)*

Täydennyslentorykmentti

By decisions made at the Air Force HQ in July and August 1940 Aviation Regiment 1 would be disbanded in a mobilization. Its squadrons (*LLv* 12, *LLv* 14 and *LLv* 16) would be subjected to various armies and *Täydennyslentorykmentti* (Supplementary Aviation Regiment) would be established for advanced training. Its main task was to train the airmen arriving from the reserves to comply with the front-line squadron requirements.

During the Continuation War mobilization *Täydennyslentorykmentti* was founded on 18 June 1941 under the command of LtCol Viljo Rekola. It had three supplementary squadrons: *T-LLv* 17, *T-LLv* 25 and *T-LLv* 35. The first one was responsible for the advanced training of bomber and reconnaissance aircrews and the other two for fighter pilots. At the beginning of the hostilities on 25 June 1941, *LLv* 34 from *LeR* 3 was subordinated to *T-LLv* 35.

The advanced training of the mobilized airmen arriving from the reserves was completed in 2–3 months and when the need clearly decreased, both *LLv* 34 and *T-LLv* 25 were annexed to *T-LLv* 35 on 1 October 1941. A month later *T-LLv* 17 was also disbanded and its parts were transferred to *LeR* 4 squadron *LLv* 46. Thus the Supplementary Aviation Regiment had shrunk to a single squadron and on 2 January 1942 became a plain headquarters, when *T-LLv* 35 was subjected to Air Force HQ.

The regiment HQ operated first from Vääksy until transferred to Lieksa on 13 January 1942. In the air force reorganization on 3 May 1942 *Täydennyslentorykmentti* was disbanded and it formed the HQ of the new Aviation Regiment 1 commanded by LtCol Viljo Rekola.

Täydennyslentolaivue 17

Capt Kyösti Kurimo was posted to command Supplementary Squadron 17 established in the mobilization on 18 June 1941. The base was set at Karvia, but already on 9 July 1941 the unit flew to Pori. The squadron was equipped with 2–4 Fokker C.Vs, 2–3 *Tuiskus*, 1–2 Fokker C.Xs, Bulldogs. *Kotkas*, Hanriots and *Pyrys*, plus half a dozen of various single types in three flights, of which one handled the twin-engine training

On 12 August 1941 Kurimo was posted to command *LLv* 10 and Maj Niilo Jusu took over the command. On 18 September 1941 the 3rd Flight (twin-engine) was transferred to Luonetjärvi and subjected to *LeR* 4 squadron *LLv* 46. It was commanded by Maj Reino Artola and when the units had no warplanes after 15 July 1941 it was formed into a training outfit of the regiment. The training flight was led by Capt Väinö Bremer.

On 1 November 1941 the whole *T-LLv* 17 was disbanded and the remaining parts were transferred to Luonetjärvi to *LLv* 46, forming its 3rd Flight headed by Capt Ragnar Laine. During its existence *T-LLv* 17 had recorded almost 1.600 flying hours

On 15 November 1941 *Lentolaivue* 48 was established under command of LtCol Raoul Harju-Jeanty. *LLv* 46 was then converted back to a bomber unit and its entire 3rd Flight was transferred to *LLv* 48, forming its 3rd Flight with Capt Laine in command.

Officer students of T-LeLv *35 refuelling a* Pyry *at Vesivehmaa on 5 June 1943. The fuel tank took 290 litres. (SA-kuva)*

Three second lieutenants of Officer Course 13 in front of a T-LeLv *35* Pyry *at Vesivehmaa on 5 June 1943. (SA-kuva)*

For twin-engine training there was at first only two Hanriots plus one Anson and a DB-3M available, until the situation improved considerably in February-March 1942 after the receipt of three Blenheims. The unit had also four *Pyrys* and one *Viima*.

T-LLv 17 and its successors 3/*LLv* 46 and 3/*LLv* 48 gave advanced training for twin-engined planes to 14 students, which fulfilled the current need. In the same period ten older pilots also received additional training in twin-engined aircraft. During August-September 1942 48 fourteen students graduated from twin-engine courses held by *LeLv* 48. Each one received 14 hours of flying training.

After the bomber purchases from Germany and the State Aircraft Factory licence production the need for advanced training of aircrews suddenly increased considerably and within *LeR* 4 Supplementary Squadron 17 (*T-LeLv* 17) was re-established on 28 November 1942, with Maj Osmo Malinen in command. The unit was based at Luonetjärvi and its later commanders can be found in the appendix. The training was carried out mainly with Blenheims and SBs.

When the Continuation War ended on 4 September 1944, *T-LeLv* 17 was disbanded a week later and its parts were transferred to *PLeLv* 46, which continued the started courses. On 26 October 1944 they were interrupted and within one month the students were demobilized, according the truce terms.

T-LeLv 17 organized nine twin-engine pilot courses, concluding all but one. The pilot courses commenced on 7 January 1943 and lasted 2–6 months. The flying time per student was 18–26 hours depending on the course. For these courses 111 students were accepted, of which 87 graduated to the front-line squadrons. The course was interrupted for 11 students and 13 were demobilized before graduation.

The advanced training for observers had mainly occurred in front-line units, but actual observer courses had not been arranged since the Winter War. When this need arose, on 7 April 1943 the first course began. In all four courses were completed and all approved 105 officers graduated. The courses lasted 2–6 months and depending on the course the flying time per student was 11–22 hours.

Lentolaivue 48 arranged in 1942 two air gunner and wireless operator courses, from which 93 students graduated to the front-line units. *T-LeLv* 17 organized a further six air gunner and wireless operator courses, the first one beginning on 7 January 1943. 105 students were approved to the one month courses and 102 of these graduated. The flying time per student was 8–30 hours depending on the course.

Täydennyslentolaivue 25

T-LLv 25 was established in the mobilization for advanced training of fighter pilots, mainly for the needs of Aviation Regiment 2. The aim was to meet the front-line requirements. Capt Eino Carlsson was appointed in command and the base was set at Ylivieska. It proved to be unsuitable and the squadron moved to Mustasaari near Vaasa on 3 July 1941.

About ten *Pyrys* and Gauntlets were received as equipment, occupying three flights. The 1st was a PY-flight, the 2nd a GT-flight and the 3rd a mixed flight. Rainy weather caused several problems with *Pyrys* and the usual strength was 3–4 planes. Additionally the unit had one *Tuisku* and a *Kotka*.

On 12 August 1941 Maj Otto Holm took over the command. Mustasaari become crowded and the unit moved to Vesivehmaa on 27 August 1941. The need for advanced training decreased strongly during the autumn and *T-LLv* 25 was disbanded on 1 October 1941 and annexed to *T-LLv* 35, also at Vesivehmaa. Thus the advanced training for fighter pilots was concentrated into one squadron.

In the theoretical fighter tactics training the intention was to follow the "Fighter Supplementary Squadron Training Programme" created in 1940 by Capt Eino Carlsson, then commanding *T-LLv* 35

A McCormick tractor pulls Blenheim BL-179 of T-LeLv 17 at Luonetjärvi on 31 March 1944. This advanced training squadron for bomber crews had on duty at most six Blenheims, of which most were fitted with dual controls. (SA-kuva)

and presently *T-LLv* 25. Carlsson was a Winter War veteran acting as a flight leader with *LLv* 24. Due to the shortage of instructors this did not come into fruition.

Some 33 students were approved to *T-LLv* 25 to receive advanced training, which per student would have been 31 flying hours. Seventeen students graduated before the unit was disbanded and rest would have done so in the next month. During its existence the squadron accumulated just over 1.000 flying hours.

Täydennyslentolaivue 35

T-LLv 35 was also founded in the mobilization, tasked with the advanced training of pilots to meet the front-line requirements of Aviation Regiment 3. At the same time *LLv* 34 was separated from *LeR* 3 and subjected to *T-LLv* 35, both now being commanded by Maj Olavi Ehrnrooth. The following squadron and flight commanders can been seen in the appendix. The new unit began operations at *LLv* 34 base Parola.

The training began with three flights: the 1st was a mixed flight flying Gamecocks, Bulldogs, I-152s and *Jaktfalkens*, the 2nd had 4–10 *Pyrys* and the 3rd flew a similar amount of Fokker D.XXIs. Additionally the squadron had a couple each of *Kotkas*, Moths and *Tuiskus*.

On 1 October 1941 the squadron moved to Vesivehmaa and in this connection both *T-LLv* 25 and *LLv* 34 were annexed to *T-LLv* 35. Having now more aircraft the 4th Flight was founded and it received 4–6 Gauntlets. The planes wore out and the flight was disbanded on 15 July 1942.

At first there was no suitable water area nearby Vesivehmaa to perform air gunnery and the FR-flight was transferred to Utti on 31 May 1942. The intended region at Haukkajärvi was also too small and the air gunnery course was held near Joensuu in July-August 1942. On 1 September 1942 the flight returned to Vesivehmaa since the north part of Vesijärvi proved to suitable for gunnery training.

Fokker D.XXI FR-152 of T-LLv *35 after suffering minor landing damage at Vesivehmaa on 16 April 1942. Typical of advanced trainers, the rudder is White. (Author's collection)*

Pyry *PY-16 of* T-LLv *25 at Mustasaari in August 1941. When this unit was disbanded on 1 October 1941, its personnel and equipment were transferred to* T-LLv *17. (Author's collection)*

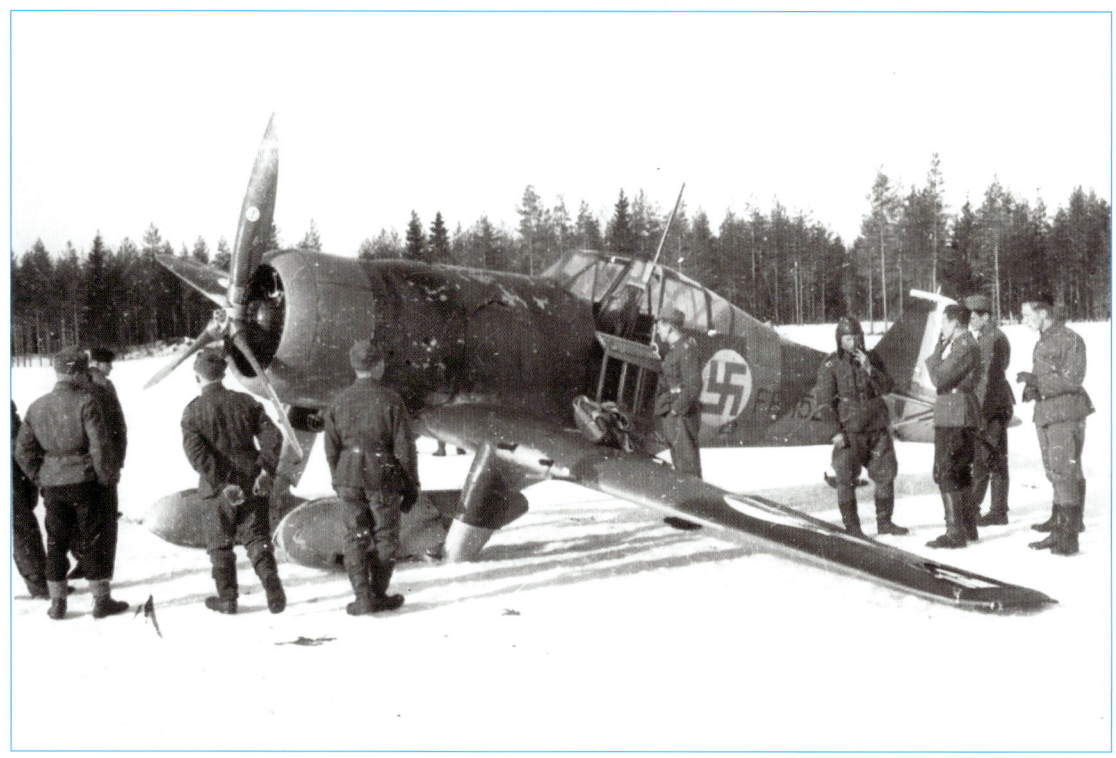

T-LeLv 35 had been subjected to the Air Force HQ from the beginning of 1942. As part of the air force reorganization, the situation also changed for *T-LeLv* 35 and from 15 September 1942 onward it was subordinated to *LeSK*.

On 1 May 1944 some internal changes were carried out in the squadron when it received 12 FIAT G. 50 fighters. They were placed in the 3rd Flight while the Fokkers now occupied the 2nd Flight.

The FIAT training took place at Utti until the whole squadron was transferred to Kauhava by 22 June 1944, away from the advancing Soviet armies.

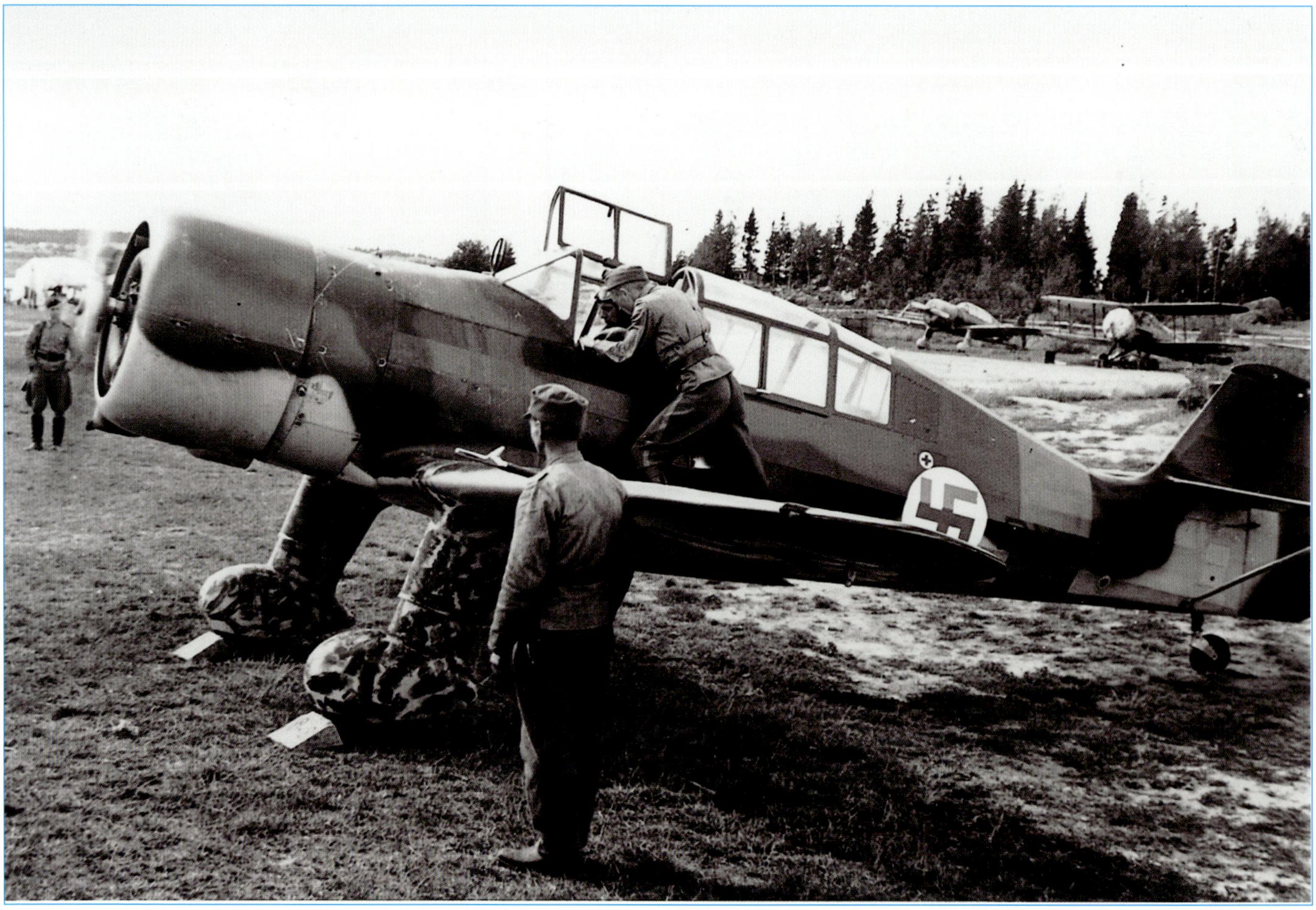

The only 4-engined aircraft of the Finnish Air Force was a de Havilland DH 89, here with civil registration OH-IPA in the Luonetjärvi hangar on 8 March 1940. It was to be serialled DH-1, but it crashed on take-off at Malmi on 27 May 1940. (SA-kuva)

The mechanics have changed the skis for wheels of this LeSK Pyry in Kauhava hangar in spring 1943. Though the skis are marked "PY-6", this in not that Pyry. (Author's collection)

The Continuation War ended on 4 September 1944 in a cease-fire and according to the truce signed two weeks later the Finnish Armed Forces were to be demobilized by 4 December 1944. In the Air Force this was done on 27 November 1944, when *T-LeLv* 35 was disbanded and annexed to the Air Fighting School.

The advanced training in the squadron had consisted of 26–48 flying hours per student. The emphasis was clearly in individual combat manoeuvres while the small formation training grew towards the end of the conflict.

On the theoretical side by far most attention was paid to flying unit tactics. Here Capt Eino Carlsson's "Fighter Supplementary Squadron Training Programme" from 1940 and updated in 1941 was effectively used. In early 1943 "Training Programme of the Supplementary Course for Air Force Officers" was introduced and also widely exploited in non-commissioned officer's advanced training. The share of tactics was raised from 14 hours to 32 hours. Physical education also played a major part.

During the war 467 students graduated from *T-LeLv* 35. Of these 265 were officers and 202 non-commissioned officers. In all the squadron performed over 23,000 flying hours.

Fokker FR-139 of T-LeLv 35 *at Vesivehmaa in October 1943. Individual planes of this unit carried markings, which could be one, two or three white bars on the rudder, white fin or white rudder. (Author's collection)*

Basic Trainers

- Letov Š-218A *Smolik*

- VL *Sääski* I and II

- De Havilland DH 60 Moth and DH 82 Tiger Moth

- VL *Viima* I, II and IIB

- Focke Wulf Fw 44 J *Stieglitz*

Letov Š-218A *Smolik*

Purchase

The Czech *Vojenskaja Tovarna Letadla Letov* designed the Š-18 primary trainer biplane in 1925. The following year improved versions, the Š-118 and Š-218, followed. The factory gave one of the latter on 2 October 1929 for trial purposes at the Aviation School. It was bought on 22 March 1930 and serialled SM-127.

After favourable experience, nine additional aircraft were bought from Czechoslovakia on 18 November 1930. A construction licence was also obtained. The Czech aircraft were in Finland by June 1931 with serials SM-128–SM-136. These were powered by 130 hp Walter NZ radials.

The State Aircraft Factory received an order for ten aircraft on 29 January 1932 and completed these by November 1933. The planes had serials SM-137–SM-146 and were fitted with 145 hp Walter Mars radial engines.

Czechoslovak-built Smolik *number two, SM-128, in the overall orange factory finish. Seen here at Kauhava in summer 1931 belonging to IlmK (Aviation School). The serial is pre-1934 style. (Finnish Air Force)*

Overall orange Smolik *SM-135 with the serial according to the 20 March 1934 directive. It is parked here at Kauhava on 19 January 1937, belonging to IlmK. (Finnish Air Force)*

On 28 March 1934 another ten plane II-series batch was ordered. It was completed by May 1935, the aircraft having serials SM-147–SM-156 and 150 hp Walter Gemma radial engines.

The last III-series batch consisted of nine aircraft ordered on 8 March 1935. The serials were SM-157–SM-165 and all but two first were powered by 160 hp Bramo Sh 14A radial engines. The aircraft were completed by October 1936.

Thus the Finnish Air Force had a total of thirty-nine Š-218A trainers in its inventory.

For most trainers, from 13 January 1939 onwards the numbers of the serial were to painted as large as could fit on the fuselage. SM-131 of ISK (Air Fighting School) stands here as an example in summer 1939. (Author's collection)

On 12 October 1939 all trainers were to be camouflaged similar to Tuisku gunnery trainers, during the next major overhaul or repair. SM-165 of ISK shows the Olive Green upper and aluminium dope lower side camouflage at Kauhava in spring 1940. (Author's collection)

A trio of LeSK (Aviation Fighting School) Smoliks at take-off from Laajalahti on 10 July 1941. They all reverted back to the overall orange finish by the 5 June 1940 order. The planes from the left are SM-164, 146 and 143. (SA-kuva)

Employment

Except for solitary aircraft all *Smoliks* saw service as elementary trainers with the Aviation School, becoming Air Fighting School on 1 January 1938. Excluding older pilots, all wartime pilots learnt to fly in *Smoliks*.

By the Winter War seven aircraft had been lost, plus two more by the time of the Continuation War. In 1941/42 only two were worn out, but in 1943 twelve more aircraft were out of service and in 1944 eleven more. At the end of the war with Russia, five aircraft remained in service. They were all flown to the depot on 6 February 1945 for storage, and later scrapped.

Colours

Except for the Caudron C.60 the standard finish for primary trainers was overall orange since 1931. All *Smoliks* had this colour as the factory finish.

After the outbreak of the Second World War an order was given on 12 October 1939 to camouflage all trainers as for the *Tuisku*, meaning Olive Green upper sides and aluminium dope lower sides. This was to be done in a major repair or overhaul.

At least fifteen *Smoliks*, serials SM-128, 131, 136, 137, 139, 141, 146, 147, 148, 152, 156, 158, 162, 164 and 165, were painted during a factory stay with olive green on top and aluminium dope on undersides.

Soon after, on 5 June 1940, new instructions were issued. The *Smoliks* were to revert to the full orange colour scheme. All to be done again at a major overhaul or repair. This all orange scheme was valid as long as the *Smoliks* flew.

Serials

Smolik SM-147 of T-LLv 35 (Supplementary Squadron) at Vesivehmaa in May 1942. It wears the regulation Orange colour with standard Yellow eastern front markings. (Olli Riekki)

Depending on the origin, different size, style and colour serials were used until 20 March 1934, when a new directive stipulated that the height of the serial was 2½ times of the arm of the fuselage swastika, much smaller than before. All serials being white in colour and the style according to the Finnish standard SFS Z.I.1.

On 13 January 1939 small letters and large numbers in white were ordered to be painted on *Smoliks*. The numbers were also introduced under both lower wings. This was done at a major overhaul or repair and was to remain throughout the careers of the *Smoliks*.

SM-159 of LeSK on skis at Kauhava in March 1943. The orthochromatic film renders the overall orange colour very dark. The markings are precisely by the book. The yellow Eastern Front markings contained a 50 cm wide rear fuselage band and the lower tips of both wings at 1/6th of the span. (Author's collection)

Smolik line-up of LeSK during an inspection held at Kauhava in July 1943. The closest machine is SM-152, which still bears the Warpaint of Olive Green and Black from its stay with LeLv 14 a year earlier. The other Smoliks are in the standard orange finish. (SA-kuva)

A working day of LeSK at Kauhava in July 1943. At left is SM-152, which was camouflaged a year earlier in the Warpaint of Black and Olive Green, for use with front-line squadron LeLv 14. (SA-kuva)

Smolik SM-162 of LeSK at Kauhava in March 1944. After twelve months all Smoliks were retired and the remaining ones put into storage. (Lars Bergman)

S/n	C/n	Delivered	Struck off charge	Remarks	Hours
SM-127	218-A1	22 Mar 1930	30 Aug 1944	W/o Sievi 7 Aug 1944	2254.10
SM-128	218-A2	9 Jun 1931	30 Sep 1943	W/o Kauhava 25 Aug 1943	2270.50
SM-129	218-A4	9 Jun 1931	20 Apr 1938	W/o Kauhava 14 Mar 1938	846.35
SM-130	218-A5	9 Jun 1931	30 Aug 1944	W/o Ylivieska 19 Jul 1944	2061.25
SM-131	218-A6	9 Jun 1931	30 Nov 1944	Into storage 27 Jul 1944	1700.40
SM-132	218-A7	9 Jun 1931	16 Feb 1944	W/o Siikakangas 25 Jan 1944	1855.25
SM-133	218-A8	9 Jun 1931	20 Sep 1941	W/o Lohtaja 22 Jul 1941	1607.05
SM-134	218-A9	9 Jun 1931	9 Aug 1943	Into storage 26 Jun 1943	2497.20
SM-135	218-A10	9 Jun 1931	28 May 1938	W/o Lappajärvi 10 Mar 1938	992.50
SM-136	218-A11	9 Jun 1931	4 Sep 1945	Into storage 6 Feb 1945	2419.40
SM-137	I/1	27 Jun 1933	20 Aug 1940	W/o Vaasa 7 Jul 1940	1364.05
SM-138	I/2	22 Aug 1933	4 Sep 1945	Into storage 6 Feb 1945	2598.45
SM-139	I/3	25 Aug 1933	7 Jul 1944	W/o Kauhava 16 Jun 1944	1923.25
SM-140	I/4	28 Aug 1933	19 Nov 1936	W/o Kauhava 15 Jun 1936	613.35
SM-141	I/5	25 Sep 1933	30 Aug 1944	W/o Ylivieska 18 Jul 1944	2390.25
SM-142	I/6	1 Oct 1933	5 Mar 1937	W/o Santahamina 16 Dec 1936	602.10
SM-143	I/7	23 Oct 1933	30 Jul 1943	W/o Kauhava 22 May 1943	2114.20
SM-144	I/8	3 Nov 1933	26 Jun 1943	W/o Kauhava 11 Aug 1942	1875
SM-145	I/9	3 Nov 1933	6 Jul 1939	W/o Lapua 31 May 1939	1047.25
SM-146	I/10	3 Nov 1933	9 Aug 1944	W/o Ylivieska 28 Jul 1944	2227.05
SM-147	II/11	25 Apr 1935	4 Sep 1945	Into storage 6 Feb 1945	2418
SM-148	II/12	2 May 1935	9 Aug 1943	W/o Kortesjärvi 11 Jun 1943	1153.15
SM-149	II/13	2 May 1935	6 Sep 1943	W/o Kauhava 4 Aug 1943	1427.45
SM-150	II/14	27 May 1935	30 Nov 1944	Into storage 17 M ar 1944	1946.45
SM-151	II/15	26 Apr 1935	30 Sep 1943	W/o Kauhava 28 Mar 1943	2212.40
SM-152	II/16	25 Apr 1935	26 Sep 1944	W/o Kortesjärvi 25 Aug 1944	2195.45
SM-153	II/17	29 Apr 1935	30 Nov 1944	Into storage 1 Aug 1944	2379.05
SM-154	II/18	3 May 1935	19 Aug 1936	W/o Kauhava 4 Jul 1936	361.05
SM-155	II/19	2 May 1935	5 Oct 1943	Into storage 19 Aug 1943	2006.35
SM-156	II/20	13 May 1935	20 Jun 1943	W/o Siikakangas 22 Mar 1943	1888.05
SM-157	III/1	3 Sep 1936	30 Nov 1943	Into storage 23 Sep 1943	2098.35
SM-158	III/2	3 Sep 1936	9 Sep 1944	W/o Halsua 25 Aug 1944	2197
SM-159	III/3	3 Oct 1936	30 Nov 1943	Into storage 23 Sep 1943	1943.20
SM-160	III/4	3 Oct 1936	30 Nov 1943	Into storage 23 Sep 1943	2023.35
SM-161	III/5	12 Oct 1936	19 Nov 1937	W/o Kauhava 21 Sep 1937	199.30
SM-162	III/6	12 Oct 1936	9 Apr 1949	Into storage 6 Feb 1945	1862.40
SM-163	III/7	10 Oct 1936	9 Apr 1949	Into storage 6 Feb 1945	1479.25
SM-164	III/8	15 Oct 1936	17 Oct 1944	W/o Pedersöre 16 Sep 1944	1860.10
SM-165	III/9	21 Oct 1936	14 Sep 1940	W/o Vaasa 10 Aug 1940	1051.35

Smolik SM-153 of LeSK in the take-off run at Kauhava in July 1943. This is a solo flight for the student occupying the rear cockpit. (SA-kuva)

Letov Š-218A Smolík, SM-130, Ilmailukoulu, Kauhava airfield, November 1932. Camouflage colours: overall Orange, serial White.

Smolík SM-130 of IlmK at Kauhava on 16 November 1932. This was a Czech-oslovak-built plane. It has the contemporary Finnish pre-1934 serial numbers. (Finnish Air Force)

Letov Š-218A Smolik, SM-154, Ilmailukoulu, Kauhava airfield, August 1935. Camouflage colours: overall Orange, serial White.

Smolik SM-154 of IlmK parked on the grass at Kauhava on 8 August 1935. The national markings and serial are strictly according to the 20 March 1934 directive. (Finnish Air Force)

Letov Š-218A Smolik, SM-164, Ilmasotakoulu, Vaasa airfield, July 1940. Camouflage colours: upper surfaces Olive Green, under surfaces aluminium dope, serial White.

SM-164 of ISK during Finnish Air Defence Association courses held at Vaasa in summer 1940. This is one of 15 Smoliks, which were camouflaged according to the 12 October 1939 order. (Lauri Volanen)

Letov Š-218A Smolik, SM-162, Lentosotakoulu, Laajalahti airfield, July 1941. Camouflage colours: upper surfaces Olive Green, under surfaces aluminium dope. Standard Eastern Front markings Yellow, serial White.

Smolik SM-162 of LeSK parked outside Laajalahti hangar in July 1941. It has retained its Olive Green top and aluminium dope undersides and wears the yellow Eastern Front recognition colours, applied on 18 June 1941. Next to it is Viima VI-3 inscribed Haijala. (Henrik Salomies)

Smolik line-up of LeSK at Laajalahti in July 1941. Planes from left are SM-163, -162 and -143. UK 13 (Officer course), which started their training here, graduated two years later. (Henrik Salomies)

Letov Š-218A Smolik, SM-158, Lentosotakoulu, Kauhava airfield, July 1943. Camouflage colours: overall Orange, standard Eastern Front markings Yellow, serial White.

Smolik SM-158 of LeSK banks to the photographer in the back seat of a Tuisku gunnery trainer over Kauhava in July 1943. (SA-kuva)

Letov Š-218A Smolik, SM-158, Lentosotakoulu, Kauhava airfield, July 1943.

Smolik SM-158 of LeSK has just taken off from Kauhava in July 1943. The overall orange primary trainer colour is clearly visible. (SA-kuva)

Letov Š-218A Smolik, SM-136, Lentosotakoulu, Kauhava airfield, May 1944. Camouflage colours: overall Orange, standard Eastern Front markings Yellow, serial White.

Six Smoliks of LeSK in a line-up at Kauhava in March 1944. The closest is SM-136 and the next SM-141. All in overall Orange and full regulation markings. (Lars Bergman)

Letov Š-218A Smolik, SM-152, Lentolaivue 14, Tiiksjärvi airfield, July 1942. Camouflage colours: upper surfaces Olive Green and Black, under surfaces Light Grey. Standard Eastern Front markings Yellow, serial Black.

SM-152 of LeLv 14 at Tiiksjärvi in July 1942, when it was used, in addition to liaison duties, as a close-range nocturnal recce plane over the immediate rear. 2/LeLv 24 (subordinated to LeLv 14) leader Capt Pauli Ervi is packing his things in front of it. (Pauli Ervi)

Smolik SM-152 at the air depot at Kuorevesi in late June 1942. It is receiving the Warpaint of Olive Green and Black, before being posted to LeLv 14, which was a front-line unit based at Tiiksjärvi. (Author's collection)

VL *Sääski* I, II, IIA and IV

Purchase

The *Sääski* (Mosquito) light biplane was a private venture designed and built in 1928 by the State Aircraft Factory staff. The air force was not interested in the type, having just acquired the licence to build de Havilland Moths. However, the prototype was bought on 25 June 1928 and coded SÄ-95. It was of wooden construction with plywood and fabric skinning and was powered by 120 hp Siemens Halske Sh 12 radial engine.

On 18 October 1929 an order was given for ten improved *Sääski* II aircraft to keep the factory running. The modifications included a 90 cm greater span and enlarged fin. All but one aircraft were completed by April 1930 and serialled SÄ-117–SÄ-125, SÄ-126 not flying until 17 June 1930.

To preserve the employment of the factory a further ten *Sääski* IIA planes were ordered on 31 October 1930. The span was further increased by 70 cm and the upper wing was fitted with slats. The planes were built by June 1931 and serialled SÄ-127–SÄ-136.

The third and last batch of twelve *Sääski* IIA versions were ordered on 15 April 1931, again to keep the factory going. These were completed by November 1931 and coded SÄ-137–SÄ-148.

In total the Finnish Air Force had thirty-three *Sääski* aircraft of four different versions.

Sääski III was only a project, with an enclosed cockpit. SÄ-120 was the only one converted to *Sääski* IV configuration in 1935 for four years. It was fitted with Paarma wing of similar structure, distinguished in having ailerons on the upper wing. Additionally a blind flying rear cockpit was built of aluminium with a fabric curtain.

The prototype, also known as the Sääski I with serial SÄ-95, after a partial repair outside the factory at Suomenlinna in early December 1929. Its service with IlmK ended on 7 August 1931 following a bad landing. (VL)

Sääski II serial SÄ-119 of MeLE at Turkinsaari outside Viipuri on 9 February 1931. The markings are factory standard for pre-March 1934 period. This particular plane served for eleven years. (Finnish Air Force)

Employment

New *Sääskis* were placed as trainers and liaison aircraft to the Aviation School at Kauhava and to all maritime units, of which the main user was *Merilentoeskaaderi* (Maritime Escadre) at Turkinsaari.

When the air stations were established on 30 June 1933, *Sääskis* went along to *Lentoasema* 4 (Air Station 4) at Turkinsaari, *LAs* 2 at Santahamina, *LAs* 6 at Viipuri and *LAs* 3 at Sortavala.

In the 1 January 1938 re-organization, *Erillinen Lentolaivue* (Detached Squadron), *Lentorykmentti* 1 (Air Regiment 1) and *Ilmasotakoulu* (Air Fighting School) became the main users.

By the Winter War *Sääskis* were concentrated into the Air Fighting School, which flew sixteen different aircraft. At the beginning of the Continuation War ten aircraft remained on duty, for liaison and as hacks of various squadrons. Only SÄ-130 survived beyond August 1941 and was struck off charge on 16 September 1943 as fully worn out.

Sääski *SÄ-117 of LAs 3 at Sortavala on 20 February 1934. This photo shows a rather common photographic phenomenon, the white circle of the national insignia looks dark in certain light conditions. (Finnish Air Force)*

Gas protection exercises at LAs 6 at Viipuri on 14 August 1936. This plane is SÄ-121, retrofitted with the slatted wing of the Sääski IIA. *The use of the type as a floatplane ended in November 1939. (Finnish Air Force)*

Colours

All *Sääskis* had factory finish of overall aluminium dope, with wing and landing gear struts in blue.

After the outbreak of the Second World War an order was given on 12 October 1939 to camouflage the trainers, similarly to the *Tuisku*. This was done during a major repair and applied to all types.

In 1940 two *Sääskis*, SÄ-125 and 146, were painted Olive Green on top and aluminium dope on bottom. Three more, SÄ-126, 130 and 133, had this scheme applied in 1941, when delivered as a hacks to front-line squadrons. These schemes remained as long as the *Sääskis* flew.

S/n	C/n	Delivered	Struck off charge	Remarks	Hours
SÄ-95	1	30 Jun 1928	16 Nov 1932	W/o Ylistaro 7 Aug 1931	412.05
SÄ-117	6	8 Mar 1930	11 Jun 1937	Into storage 14 Apr 1937	1166.30
SÄ-118	7	12 Mar 1930	9 Jan 1943	W/o Uukuniemi 21 Aug 1941	1616.30
SÄ-119	8	14 Mar 1930	12 Nov 1937	Worn out	1325.40
SÄ-120	9	2 Apr 1930	27 Sep 1941	W/o Värtsilä 20 Aug 1941	1811.55
SÄ-121	10	2 Apr 1930	9 Jan 1943	W/o Tohmajärvi 23 Jul 1941	1408.10
SÄ-122	11	11 Apr 1930	16 Jul 1936	W/o Karhusuo 13 Feb 1936	993.25
SÄ-123	12	12 Apr 1930	22 Jul 1937	W/o Turkinsaari 3 Jun 1937	1191.35
SÄ-124	13	22 Apr 1930	7 Jun 1937	Into storage 14 Apr 1937	1056.10
SÄ-125	14	30 Apr 1930	9 Jan 1943	W/o Kitee 21 Aug 1941	1297.30
SÄ-126	15	17 Jun 1930	9 Jan 1943	Into storage 30 Jul 1941	1641.50
SÄ-127	17	2 Apr 1931	14 Jun 1937	Into storage 20 May 1937	1162.15
SÄ-128	18	2 Apr 1931	9 Apr 1940	W/o Kauhava 7 Jan 1940	1477.20
SÄ-129	19	15 Apr 1931	2 Nov 1940	W/o Oravainen 28 Apr 1940	1490.55
SÄ-130	20	22 Apr 1931	16 Sep 1943	W/o Kirvu 10 Sep 1942	1484.10
SÄ-131	21	22 Apr 1931	23 Dec 1936	Into storage 15 Jun 1936	825.40
SÄ-132	22	23 Apr 1931	2 Nov 1940	W/o Keuruu 12 Jun 1940	1499
SÄ-133	23	22 May 1931	20 Sep 1941	W/o Kaurila 1 Aug 1941	1449.10
SÄ-134	24	9 Jun 1931	16 Oct 1940	Worn out	1165
SÄ-135	25	11 Jun 1931	27 Sep 1938	W/o Turkinsaari 30 Jun 1938	1403.25
SÄ-136	26	18 Jun 1931	11 Oct 1941	W/o Lunkula 12 Aug 1941	1136.20
SÄ-137	IIa/11	3 May 1932	21 May 1940	W/o Mänkijärvi 26 Mar 1940	1470.05
SÄ-138	IIa/12	17 Feb 1932	9 Jan 1943	W/o Salmi 25 Aug 1941	1027.15
SÄ-139	IIa/13	17 Feb 1932	29 Nov 1938	W/o Sortavala 11 Oct 1938	1174.20
SÄ-140	IIa/14	23 Feb 1932	7 Mar 1933	W/o Sortavala 28 Jan 1933	336.50
SÄ-141	IIa/15	22 Feb 1932	7 Mar 1933	W/o Sortavala 28 Jan 1933	321.05
SÄ-142	IIa/16	29 Feb 1932	24 Sep 1932	W/o Johannes 1 Sep 1932	264.20
SÄ-143	IIa/17	22 Jun 1932	22 Jan 1941	W/o Kauhava 4 Dec 1940	1168.10
SÄ-144	IIa/18	4 Mar 1932	28 Jun 1933	W/o Santahamina 9 Dec 1932	210.05
SÄ-145	IIa/19	13 Mar 1932	2 Nov 1940	Worn out	1603.35
SÄ-146	IIa/20	24 Mar 1932	18 Aug 1941	W/o Mikkeli 22 Jun 1941	1440.35
SÄ-147	IIa/21	22 Jun 1932	3 Sep 1937	Worn out	889.15
SÄ-148	IIa/22	7 Apr 1932	12 Feb 1940	W/o Mänkijärvi 23 Nov 1939	1290.30

Sääski SÄ-130 was a hack of LeLv 26. It is seen here at Immola shortly before 10 September 1942, when it was damaged in a bad landing beyond repair. The camouflage is regulation Olive Green top and aluminium dope undersides. (Author's collection)

VL Sääski IIA, SÄ-131 of Lentoasema 3, Sortavala air station, February 1934. Aluminium dope finish overall, tail stripes Bue, serial Black.

Sääski SÄ-131 of LAs 3 parked at Sortavala on 20 February 1934. The front fuselage has the unit's Karelian Sword emblem. The blue tail stripes indicate the personal plane of the commander, Maj Erik Stenbäck. (Finnish Air Force

44

VL Sääski IIA SÄ-145, Ilmasotakoulu, Kauhava airfield, October 1940. Aluminium dope finish overall, serial and tail number Black.

Sääski SÄ-145 of ISK at Kauhava before being put into storage on 2 November 1940 as totally worn out. For in-flight recognition the tail bore the last two digits of the serial number. (Author's collection)

VL Sääski II, SÄ-125 of Lentorykmentti 3, Siikakangas airfield, April 1941. Camouflage colours: Olive Green upper surfaces and aluminium dope under surfaces, serial and tail number Black.

Sääski SÄ-125 was the liaison aircraft of LeR 3 headquarters. It is seen here at Siikakangas after delivery, which took place on 10 April 1941. The Olive Green upper colour was ordered on 12 October 1939. The tail number 25 was a carry-over from the earlier training era. (Author's collection)

VL Sääski IIA, SÄ-130 of Lentorykmentti 3, Utti airfield, December 1941. Camouflage colours: Olive Green brush strokes over aluminium dope finish, standard Eastern Front markings Yellow, serial Black.

Another liaison aircraft of LeR 3 was SÄ-130, here on a visit to Utti in December 1941. The aluminium dope overall finish was camouflaged temporarily by brush strokes of Olive Green paint. A couple of months later it received a solid camouflage. (Author's collection)

De Havilland D.H. 60X Moth and D.H.82 Tiger Moth

Purchase

On 31 March 1928 the State Aircraft Factory obtained a licence from de Havilland to build the export version of the Moth. The I-series of eight aircraft was completed in February and March 1929, the planes receiving serials MO-96–MO-103. These were powered by a 80 hp Cirrus I in-line engine.

Twelve months later the factory constructed the II-series of eleven aircraft, serialled MO-106–MO-116. These were powered by 115 hp Cirrus Hermes II in-line engines.

In 1939 one more Moth was received as a gift and three were requisitioned. These were serialled MO-93, 94, 103 (again) and 104. Thus the Finnish Air force had twenty-three Moths in total.

The only D.H. 82 Tiger Moth was interned on 8 June 1940, having escaped from the German invasion of Norway. It was serialled MO-159 after the original Norwegian number.

The first Finnish licence-built D.H. 60X Moth, serial MO-96, ready for its maiden flight from the ice at Suomenlinna, outside Helsinki, on 30 January 1929. The markings are fully factory standard of the period. The first batch was fitted an 80 hp Cirrus I in-line engine. (VL)

During the first decade of service, the Moths were usually fitted with floats during the open water season. Here from left two Moths, MO-107 and 112, and Sääski SÄ-130 of MeLAs on the bridges at Santahamina on 16 May 1931. (Finnish Air Force)

Employment

New Moths were placed as trainers and liaison aircaft, fitted with wheels, skis and floats to all units: Aviation School (*IlmK*) at Kauhava, Maritime Escadre (*MeLE*) at Turkinsaari, 1st Detached Maritime Squadron (1. *ErMeLL*) at Viipuri, 2nd Detached Maritime Squadron (2. *ErMeLL*) as Sortavala, Detached Land Squadron (*ErMLL*) at Suur-Merijoki, Land Escadre (*MLE*) at Utti and Land Aviation School (*MaalK*) at Santahamina.

The air stations were formed on 30 June 1933 and Moths served as land planes at Kauhava's Aviation School, Utti's Air Station 1 and Suur-Merijoki's Air Station 5.

On 1 January 1938 the air station were converted to air regiments and, in addition to the Air Fighting School (*ISK*), Moths served as liaison planes in the squadrons of all regiments.

During and after the Winter War twelve Moths were on duty, half with the Air Fighting School and the rest for liaison in various squadrons. The Continuation War saw just seven Moths flying for liaison and as squadron hacks. The last flight was conducted by MO-112 on 7 November 1944, when taken to storage at the depot.

Tiger Moth MO-159 flew for three weeks in June 1942 with *LeLv* 14 and from August 1943 as a hack with *LeLv* 26 and 24, until its crash on 17 May 1944.

Moth MO-100 outside Veljekset Karhumäki Oy (Karhumäki Brothers Ltd) aircraft factory at Keljo near Jyväskylä after a full overhaul in July 1934. The markings comply with the recent 20 March 1934 regulations. (Karhumäki))

Moth MO-104 as the hack of LeLv 26 after arrival at Kilpasilta on 27 April 1942. It had the regulation camouflage of Olive Green upper and aluminium dope lower surfaces. The second batch engine was a 115 hp Cirrus Hermes II in-line. (Olli Riekki)

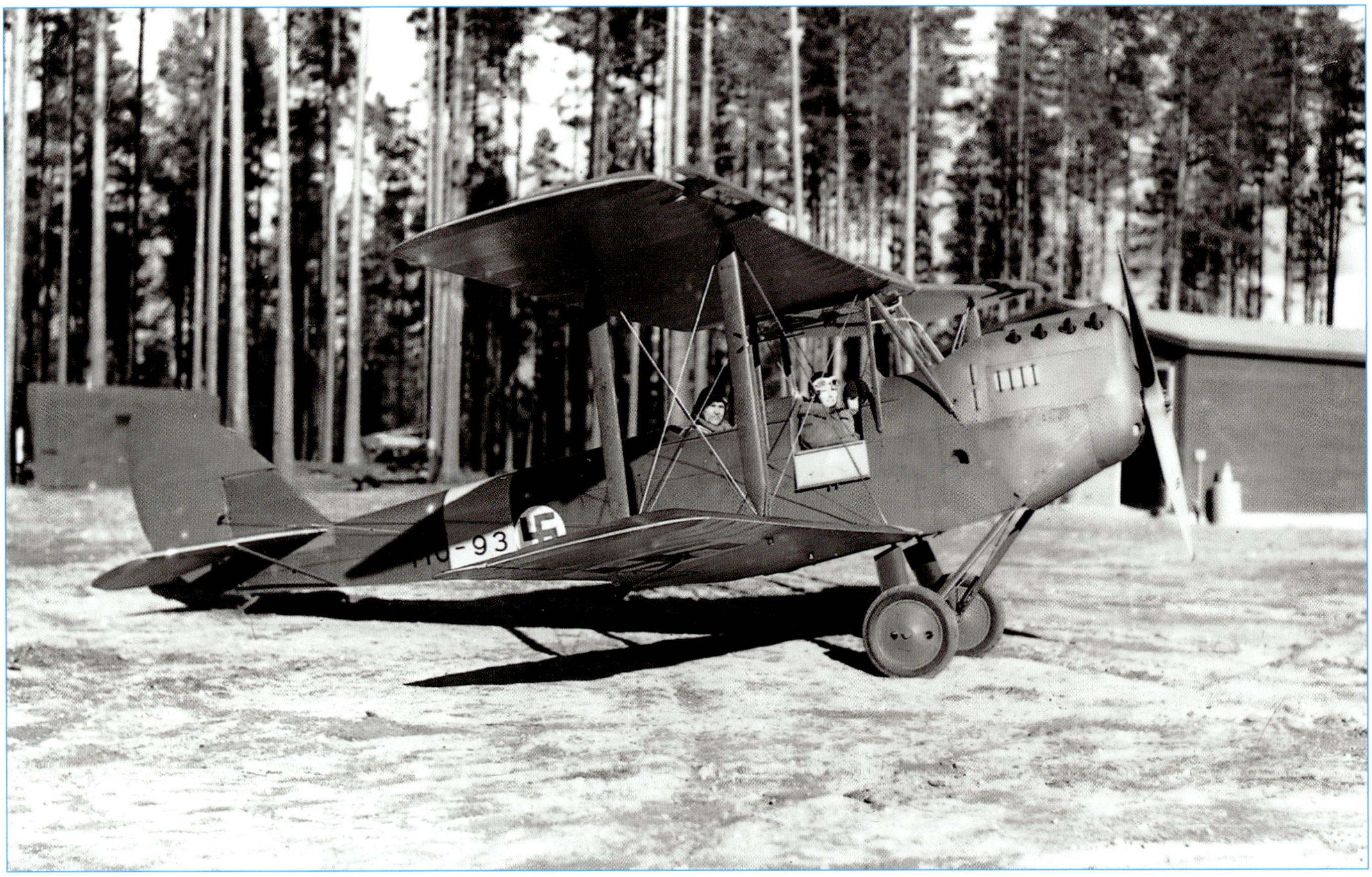

Moth MO-93 was used for liaison by LeLv *30 and is seen here before take-off at Suulajärvi in June 1942. The plane shows a pristine camouflage of Olive Green. (Aulis Bremer)*

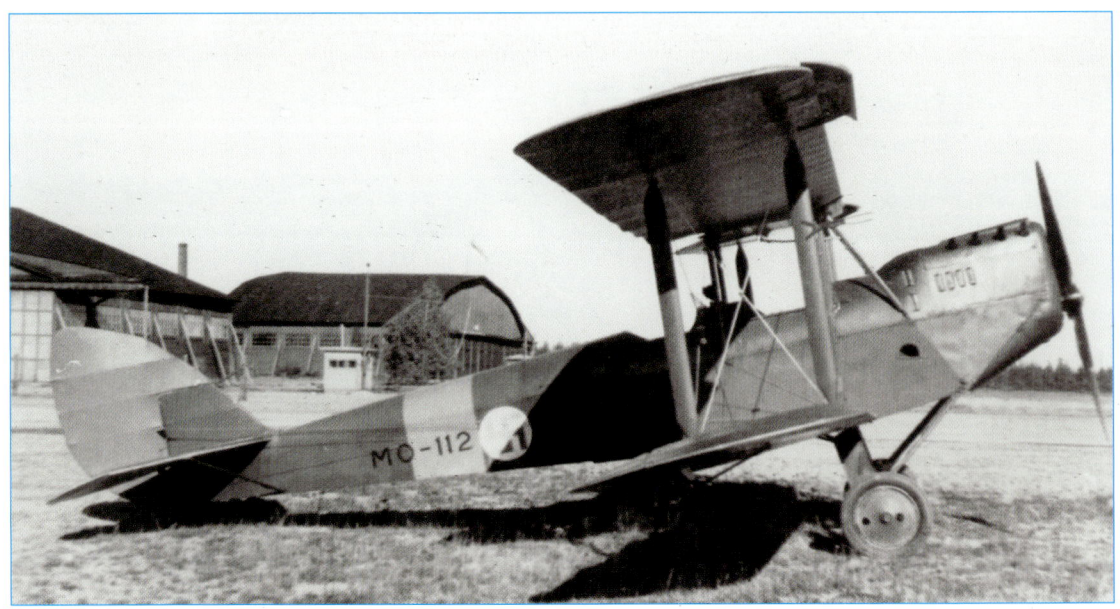

Colours

All Moths were originally painted in overall aluminium dope at the factory, with Blue wing, tail-plane and landing gear struts.

After the outbreak of the Second World War an order was given on 12 October 1939 to camouflage the trainers similarly to the *Tuisku*. This was done in a major repair and applied to all types.

In 1940 six Moths, MO-94, 103, 104, 106, 108 and 114, were painted Olive Green on top and aluminium dope on the bottom. In 1941 a further four Moths, MO-93, 102, 109 and 112, received the same camouflage. These schemes was valid as long as the Moths flew.

Serials

Large and different serials were used until 20 March 1934, when a directive stipulated that the height of the serial was 2½ times of the arm of the fuselage swastika, much smaller than before.

Moth MO-112 as the hack of LeLv *34, seen at Utti in June 1943. This machine performed the last flight of the type in the Finnish Air Force on 7 November 1944. (Eino Ritaranta collection)*

S/n	C/n	Delivered	Struck off charge	Remarks	Hours
MO-93	VK 1	20 Nov 1939	9 Aug 1944	W/o Hirvas on 9 Apr 1944	1500
MO-94	VK 2	24 Oct 1939	30 Aug 1941	W/o Naarajärvi 27 Jun 1941	689.10
MO-96	1	2 Feb 1929	18 Mar 1937	W/o Kauhava 12 Feb 1937	901.15
MO-97	2	28 Feb 1929	5 Jan 1931	W/o Viipuri 28 Sep 1930	292.20
MO-98	3	28 Feb 1929	8 May 1931	W/o Viipuri 13 Apr 1931	266.35
MO-99	4	15 Mar 1929	2 Jan 1933	W/o Alahärmä 12 Dec 1932	375.05
MO-100	5	6 Mar 1929	10 Jun 1937	W/o Kauhava 1 Apr 1937	778.25
MO-101	6	12 Mar 1929	22 Jul 1937	W/o Utti 9 Jun 1939	1002.35
MO-102	7	19 Mar 1929	28 Jun 1944	W/o Kangasniemi 30 May 1944	1176.40
MO-103	8	22 Mar 1929	14 May 1929	Into storage and then sold	21.50
MO-103	VK 3	19 Dec 1940	10 Oct 1942	W/o Jyrkänmäki 10 Jul 1942	113.10
MO-104	DH 992	19 Dec 1940	26 Sep 1944	W/o Lappee 24 Jul 1944	398.30
MO-106	11	19 Feb 1930	9 Apr 1940	W/o Lappajärvi 29 Jan 1940	1502.45
MO-107	12	22 Feb 1930	17 Jun 1932	W/o Helsinki 26 Mar 1932	278
MO-108	13	22 Feb 1930	13 Apr 1944	W/o Suulajärvi 10 Mar 1944	1641.05
MO-109	14	25 Feb 1930	16 Mar 1943	W/o Nurmoila 7 Jan 1943	1148.25
MO-110	15	7 Mar 1930	1 Nov 1938	W/o Puumala 24 Sep 1938	1200.40
MO-111	16	8 Mar 1930	3 Jan 1940	W/o Immola 1 Dec 1939	1132.25
MO-112	17	7 Mar 1930	18 Dec 1944	Into storage 7 Nov 1944	2059.20
MO-113	18	22 Mar 1930	21 Jun 1940	W/o Mänkijärvi 9 May 1940	1377.30
MO-114	19	30 Mar 1930	2 Jul 1940	W/o Mänkijärvi 4 May 1940	1170.25
MO-115	20	10 Apr 1930	14 Nov 1930	W/o Johannes 16 Jun 1930	32.50
MO-116	21	29 Apr 1930	1 Nov 1938	W/o Ruokolahti 20 Sep 1938	1016.35
MO-159	Kjeller 165	9 Jun 1940	30 Aug 1944	W/o Suulajärvi 17 May 1944	117.15

D,H. 82 Tiger Moth serial MO-159 of LeLv 26 parked at Kilpasilta, where it arrived on 18 August 1943. It served as for liaison with this squadron until 4 April 1944, when handed over to HLeLv 24, only to crash seven weeks later. (Olli Riekki)

De Havilland D.H. 60X Moth, MO-110, Erillinen Maalentolaivue,
Suur-Merijoki, November 1930. Aluminium dope finish overall, serial Black.

MO-110 arrived to ErMLL at Suur-Merijoki on 6 November 1930. The Moths were used in the in-unit training in most outfits.
(Finnish Air Force)

De Havilland D.H. 60X Moth, MO-110, Erillinen Maalentolaivue,
Suur-Merijoki, November 1930.

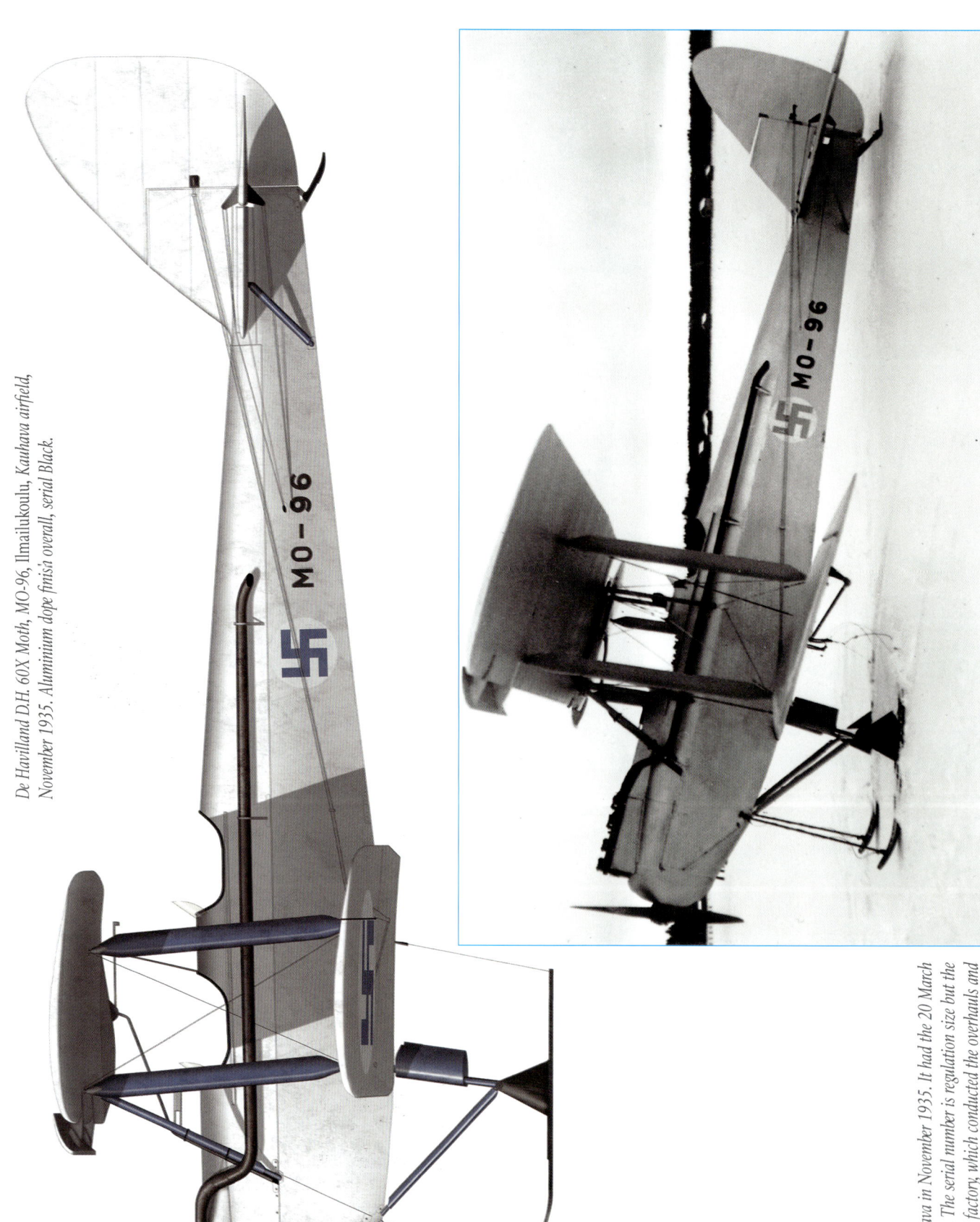

De Havilland D.H. 60X Moth, MO-96, Ilmailukoulu, Kauhava airfield, November 1935. Aluminium dope finish overall, serial Black.

Moth MO-96 of IlmK at Kauhava in November 1935. It had the 20 March 1934 regulation national insignia. The serial number is regulation size but the style is typical of the Karhumäki factory, which conducted the overhauls and repairs of the type. (Author's collection)

De Havilland D.H. 60X Moth, MO-104, Lentolaivue 26, Kilpasilta, April 1943. Camouflage colours: Olive Green upper and aluminium dope under surfaces, standard Eastern Front markings Yellow, serial Black.

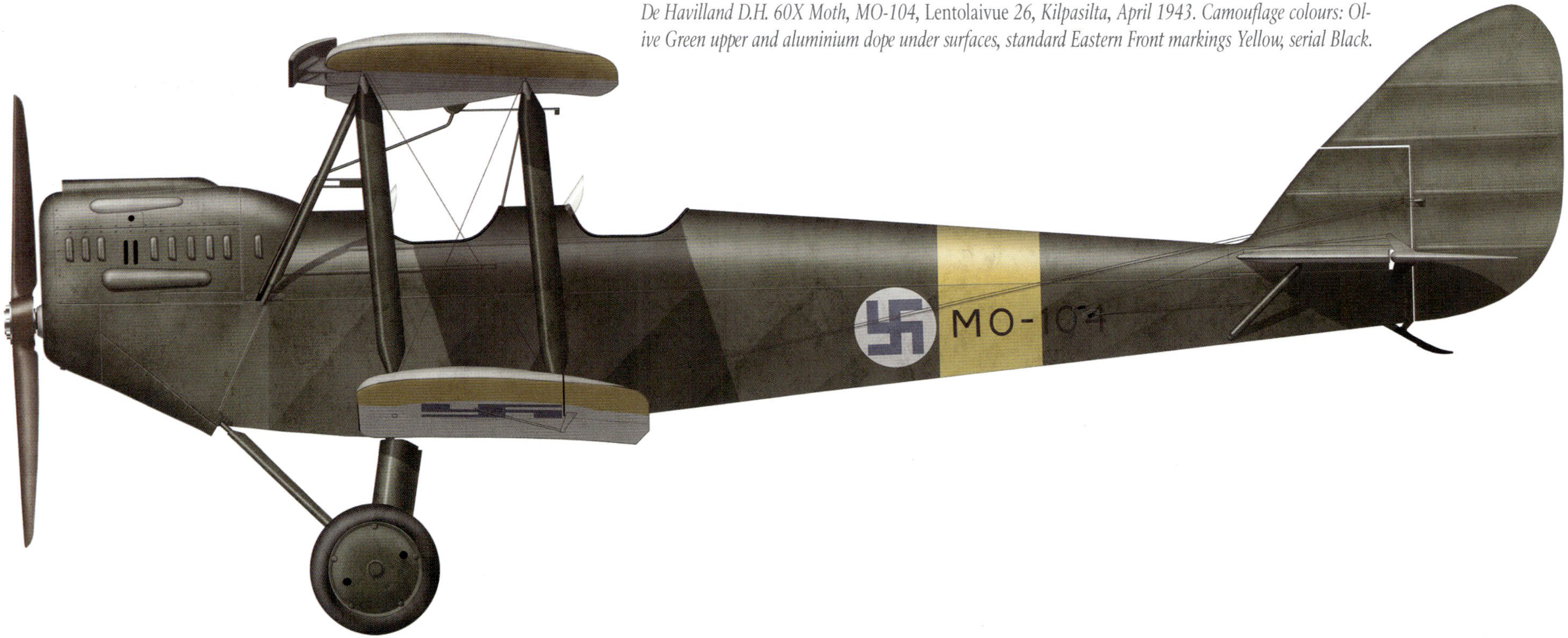

MO-104 was the hack of LeLv 26 and it is seen here at Kilpasilta in spring and summer 1943, before being put temporarily into storage on 18 August 1943. (Olli Riekki)

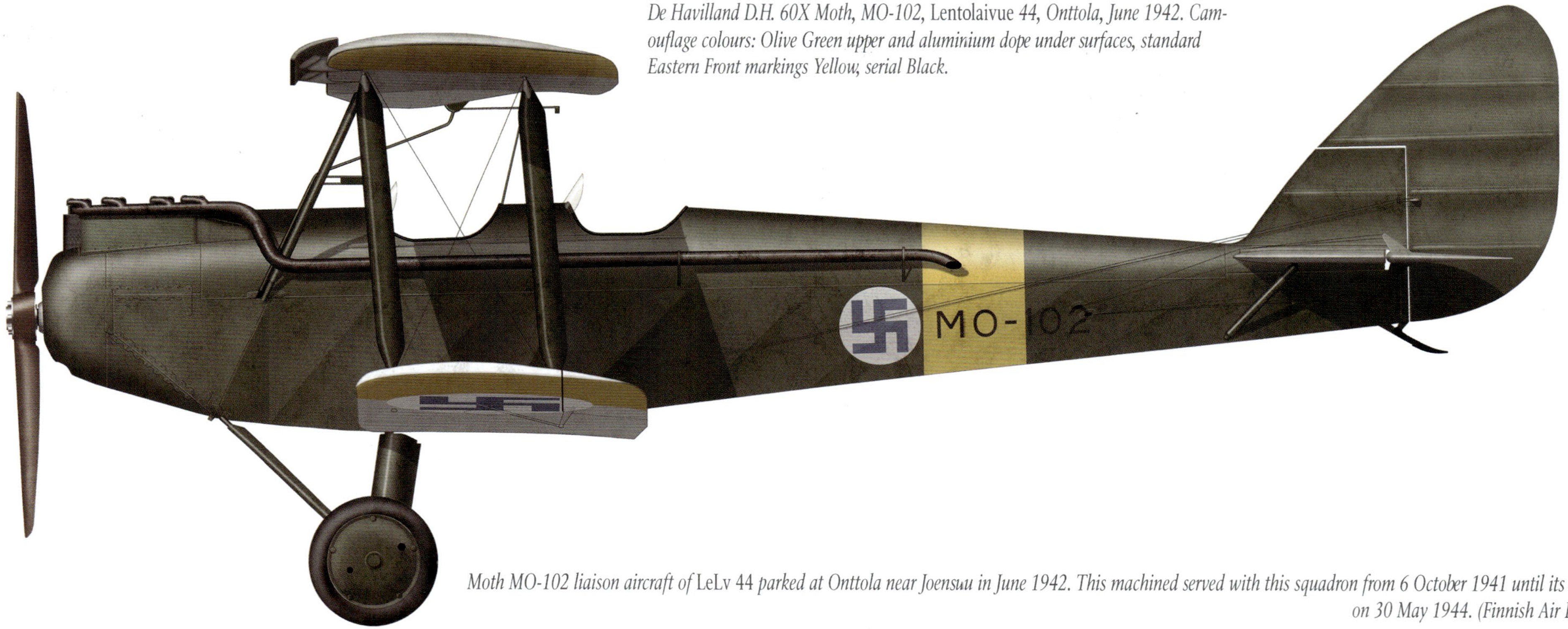

De Havilland D.H. 60X Moth, MO-102, Lentolaivue 44, Onttola, June 1942. Camouflage colours: Olive Green upper and aluminium dope under surfaces, standard Eastern Front markings Yellow, serial Black.

Moth MO-102 liaison aircraft of LeLv 44 parked at Onttola near Joensuu in June 1942. This machined served with this squadron from 6 October 1941 until its crash on 30 May 1944. (Finnish Air Force)

De Havilland D.H. 60X Moth, MO-102, Lentolaivue 44, Onttola, June 1942.

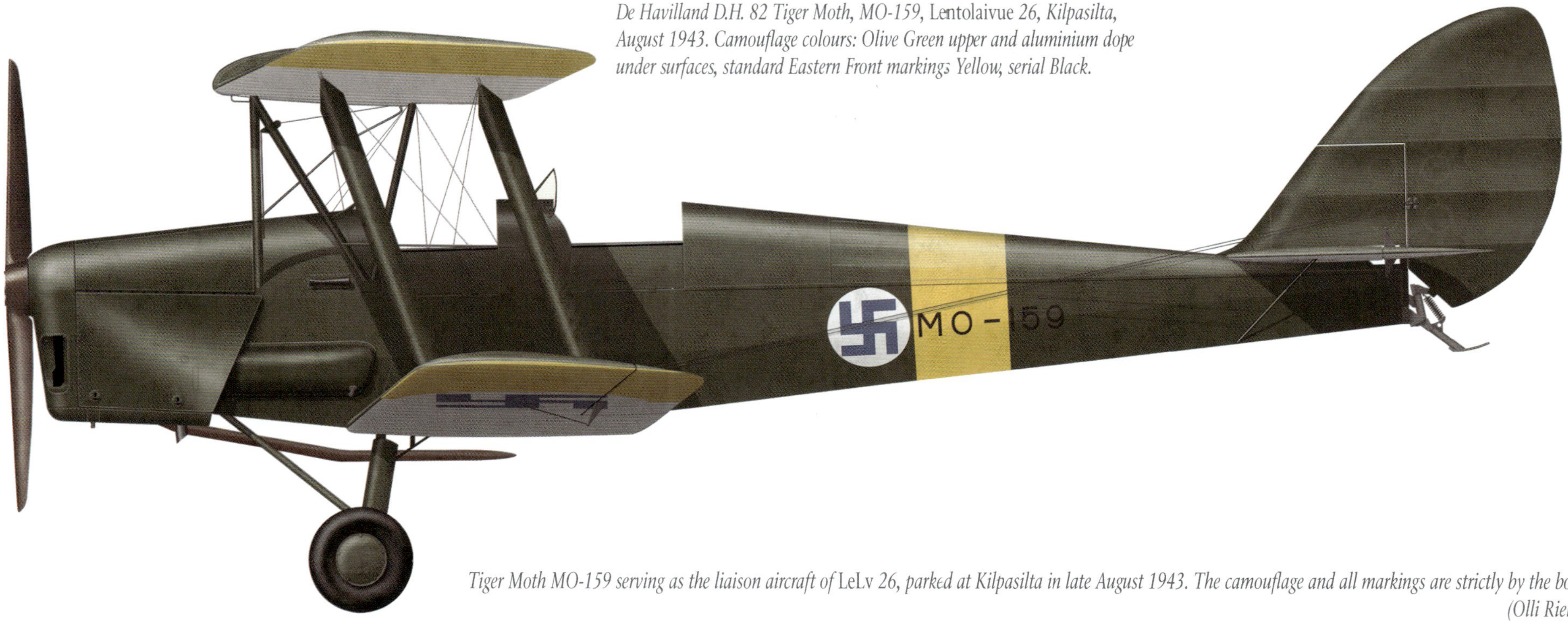

De Havilland DH. 82 Tiger Moth, MO-159, Lentolaivue 26, Kilpasilta, August 1943. Camouflage colours: Olive Green upper and aluminium dope under surfaces, standard Eastern Front markings Yellow, serial Black.

Tiger Moth serving as the liaison aircraft of LeLv 26, parked at Kilpasilta in late August 1943. The camouflage and all markings are strictly by the book.
(Olli Riekki)

Tiger Moth MO-159 serving as the liaison aircraft of LeLv 26, parked at Kilpasilta in late August 1943.

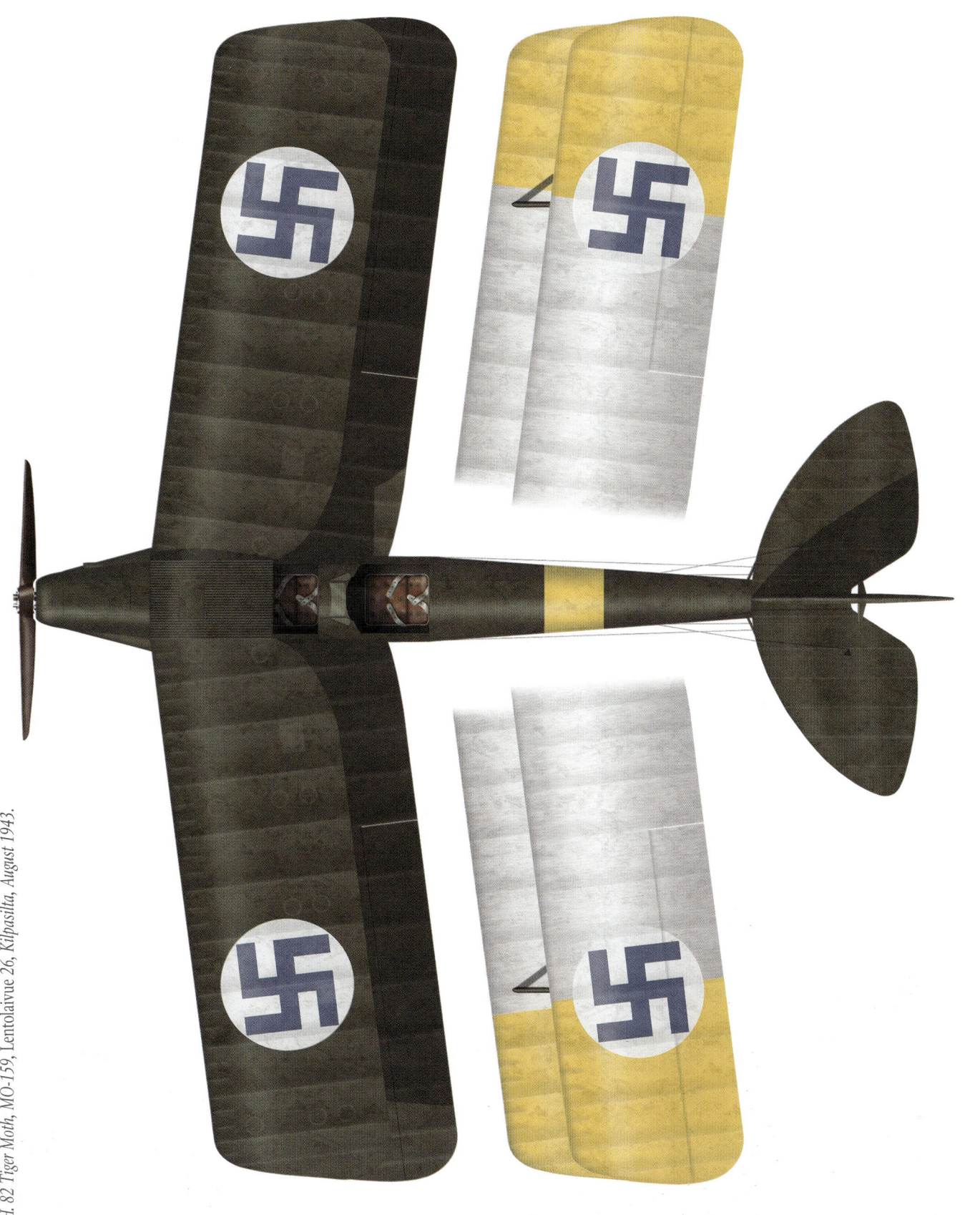

VL *Viima* I, II and IIB

Purchase

To replace the *Smoliks* the air force decided to obtain a domestic type. On 7 February 1935 the factory received on order for one elementary trainer prototype. Arvo Ylinen chose the same construction and configuration as the *Tuisku*, to be powered by 140 hp Siemens Halske Sh 14 radial engine.

The prototype was named *Viima* I (Draught) and serialled VI-1. The first flight was performed on 11 January 1936. Though good, the flying characteristics were not quite to the specifications.

On 24 April 1937 the second prototype VI-2 was ordered. It was otherwise similar but somewhat smaller and made its maiden flight on 12 October 1937. All specifications were met and on 27 June 1938 an order for 22 aircraft was placed.

The Viima *first prototype, VI-1, at the State Aircraft Factory at Tampere in June 1936. It displays nicely the overall trainer orange colour. This prototype was about 10% bigger than the subsequent aircraft. (VL)*

Viima *second and slightly smaller prototype, VI-2, outside the factory at Tampere in May 1938. The overall colour is the specified orange. All markings comply with the 20 March 1934 regulations. (VL)*

The aircraft were completed between July and November 1939 with serials VI-3–VI-22. The non-commissioned officers raised the funds for VI-3, which was inscribed "*Haijala*". The elementary school teachers sponsored VI-4, which was inscribed "*Kokko*". Two additional aircraft were built for the civil air defence association under registrations OH-ILM and ILN.

OH-ILN was transferred to the air force in October 1939 and serialled VI-23. In February 1953 OH-VKJ was bought from Karhumäki and serialled VI-40. It was equipped with an enclosed cockpit and Cirrus Major in-line engine. The enclosed cockpit became a standard fitting after February 1954. Thus the Finnish Air Force had 24 *Viimas* in its inventory.

Employment

New *Viimas* were mainly deployed to the Air Fighting School. By the Continuation War only one had been lost and the Aviation Fighting School remained as the main user. Individual *Viimas* served in liaison duties and as hacks in several squadrons.

After the war in 1945 nineteen aircraft were still in service. The Air Fighting School was the main employer and all other air force units had *Viimas* as liaisons and hacks.

Eight aircraft were lost in peacetime. On 31 May 1960 ten *Viimas* were sold in an auction to air clubs and eight eventually entered the civil register.

Four *Viimas* remained, though three were lost in summer 1960. VI-21 flew for the last time on 17 August 1962 and was later sold and entered the civil aviation register.

All Viimas *in for main overhaul or repair had the 5 June 1940 paint scheme: Olive Green fuselage and Orange wings and tail. VI-5 of LeSK demonstrates this at Laajalahti camp in July 1941, plus the Yellow Eastern Front markings introduced on 18 June 1941. (Henrik Salomies)*

Viima *VI-11 of LeSK* parked on the ramp at Kauhava in July 1941, wearing the regulation Green and Orange colours. The large number of the serial was ordered on 13 May 1939 for Viimas. (Author's collection)

Viima *VI-16 of LeSK taxis* at Kauhava in August 1941. Number 16 of the serial is clearly visible under the lower port wing. The order stated that this number should be under both wings, but it was occasionally only under the port one. (Author's collection)

The Viima first prototype flew many years in active service. Here is *VI-1 of LeSK at Lappajärvi in March 1942,* wearing the regulation markings and colours of Green and Orange. (Olli Riekki)

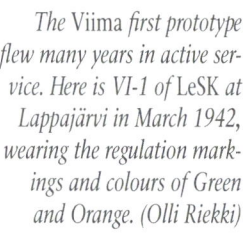

Viima VI-11 was the liaison aircraft of LeLv 26, parked here at Kilpasilta in May 1943. The closeness of the front on the Karelian Isthmus caused the painting of camouflage on the plane, upper sides in Olive Green and lower sides in DN-colour. (Olli Riekki)

Colours

The intended factory finish for the *Viimas* was the regulation primary trainer orange.

After the outbreak of the Second World War an order was given on 12 October 1939 to camouflage the trainers similarly to the *Tuisku*. This was done in a major repair and applied to all types.

At this moment the last five *Viimas* were still under construction and thus VI-18, 19, 20, 21 and 22 were painted olive green on top and aluminium dope on bottom. Also VI-1 and 2 received this camouflage in early 1940.

On 5 June 1940 new instructions were issued. The fuselage of *Viimas* was to be painted Olive Green while the wings and tail were to be Orange. These schemes were valid as long as the *Viimas* flew.

In the middle of the Continuation War *LeLv* 12, working on the Olonets front, received three *Viimas*, VI-1, 13 and 14 for nocturnal close reconnaissance role. These planes received on 13 October 1942 the more suitable Warpaint of Black and Olive Green tops with DN colour undersides. This camouflage was carried until January 1944, when all three had become damaged in flying accidents.

Viima VI-1 liaison of LeLv 12 at Voronpää shortly before its take-off failure on 19 August 1943. Typical of front-line squadrons the hacks also received the Warpaint of Olive Green and Black with DN-colour undersides. (Finish Air Force Museum)

Viima *VI-15* at Luonetjärvi on 26 October 1943, having just become the liaison craft of LeLv 46. All markings are by the book and standard trainer colours of Green and Orange. This and the next VI-12 have the enlarged fin fitted for better diving properties. (Finnish Air Force)

Viima *VI-12* of LeSK at Laajalahti camp in August 1944, in regulation markings. The enlarged fin fitted here was ordered on 11 July 1941, to be installed at the next major overhaul or repair. (Finnish Air Force)

Serials

Large and different serials were used until 20 March 1934, when a directive stipulated that the height of the serial was 2½ times of the arm of the fuselage swastika, much smaller than before.

On 13 May 1939 small letters and large numbers in white were ordered to be painted on *Viimas*. The numbers were also introduced under both lower wings. This also transferred the national marking from the fuselage to the rudder. This was done at a major overhaul or repair and was to remain throughout the careers of *Viimas*.

S/n	C/n	Delivered	Struck off charge	Remarks	Hours
VI-1		22 Feb 1936	2 Jul 1953	Into storage 11 Oct 1949	2406
VI-2		17 Nov 1937	31 May 1960	Into storage 13 Oct 1959	5540.45
VI-3	I/1	1 Aug 1939	31 May 1960	Into storage 23 Jun 1959	4004.35
VI-4	I/2	17 Aug 1939	22 Jun 1950	W/o Kauhava 18 Jul 1949	1869
VI-5	I/3	18 Sep 1939	16 Feb 1954	W/o Kauhava 18 Jan 1954	2607.40
VI-6	I/4	7 Sep 1939	31 May 1960	Into storage 25 Feb 1960	3875.35
VI-7	I/5	25 Aug 1939	31 May 1960	Into storage 28 Jan 1959	4324.10
VI-8	I/6	18 Sep 1939	2 Nov 1943	W/o Kausala 4 Oct 1943	740.50
VI-9	I/7	21 Sep 1939	7 Jun 1957	W/o Lappajärvi 15 Apr 1957	3944.05
VI-10	I/8	21 Sep 1939	9 Apr 1940	W/o Halsua 26 Jan 1940	215.25
VI-11	I/9	27 Sep 1939	16 Sep 1943	W/o Jääski 27 Jul 1943	1288.20
VI-12	I/10	27 Sep 1939	31 May 1960	Into storage 11 Sep 1959	4528.05
VI-13	I/11	2 Oct 1939	31 May 1960	Into storage 8 Feb 1960	4124.15
VI-14	I/12	10 Oct 1939	20 Sep 1947	W/o Muuruvesi 6 Aug 1947	1393.55
VI-15	I/13	18 Oct 1939	31 May 1960	Into storage 13 May 1960	4564.35
VI-16	I/14	18 Oct 1939	31 May 1960	Into storage 13 May 1960	4117.55
VI-17	I/15	18 Oct 1939	31 May 1960	Into storage 23 Mar 1959	4435.50
VI-18	I/16	25 Oct 1939	20 Sep 1941	W/o Kauhava 14 Aug 1941	881
VI-19	I/17	6 Nov 1939	16 Apr 1959	W/o Luumäki 12 Jul 1957	3768.45
VI-20	I/18	8 Nov 1939	31 May 1960	Into storage 17 Jun 1959	3848.45
VI-21	I/19	18 Nov 1939	10 Apr 1963	Into storage 17 Aug 1962	5571.15
VI-22	I/20	18 Dec 1939	28 Oct 1960	W/o Pori 9 Jun 1960	4442.25
VI-23	OH-ILN	19 Oct 1939	31 Oct 1960	W/o Sippola 4 Jul 1960	4094.25
VI-40	OH-VKJ	19 Feb 1953	31 Oct 1960	W/o Luonetjärvi 24 Jun 1960	2026.20

The end of the line was Viima IIB VI-40, seen here at Tampere in autumn 1954. It was fitted with a 130 hp Gipsy Major in-line engine and enclosed cockpit. The latter became standard on Viimas from 9 February 1954 onwards. (VL)

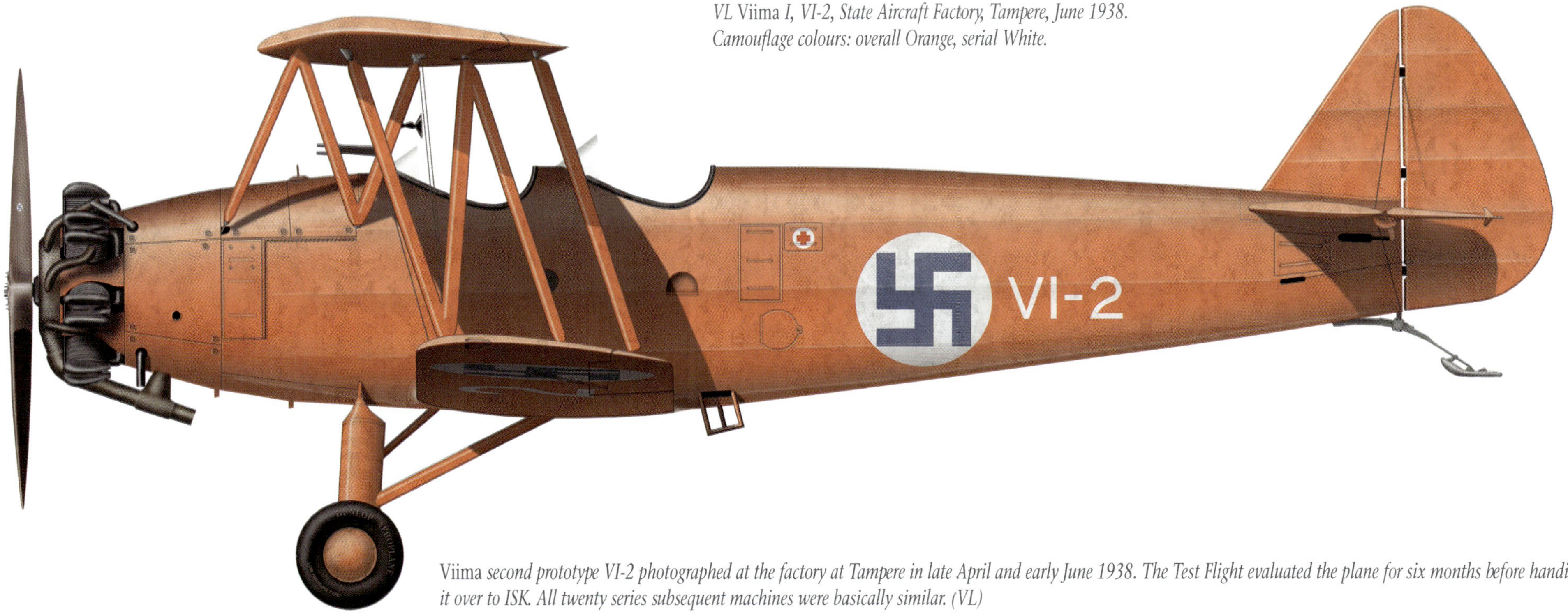

VL Viima I, VI-2, State Aircraft Factory, Tampere, June 1938. Camouflage colours: overall Orange, serial White.

Viima second prototype VI-2 photographed at the factory at Tampere in late April and early June 1938. The Test Flight evaluated the plane for six months before handing it over to ISK. All twenty series subsequent machines were basically similar. (VL)

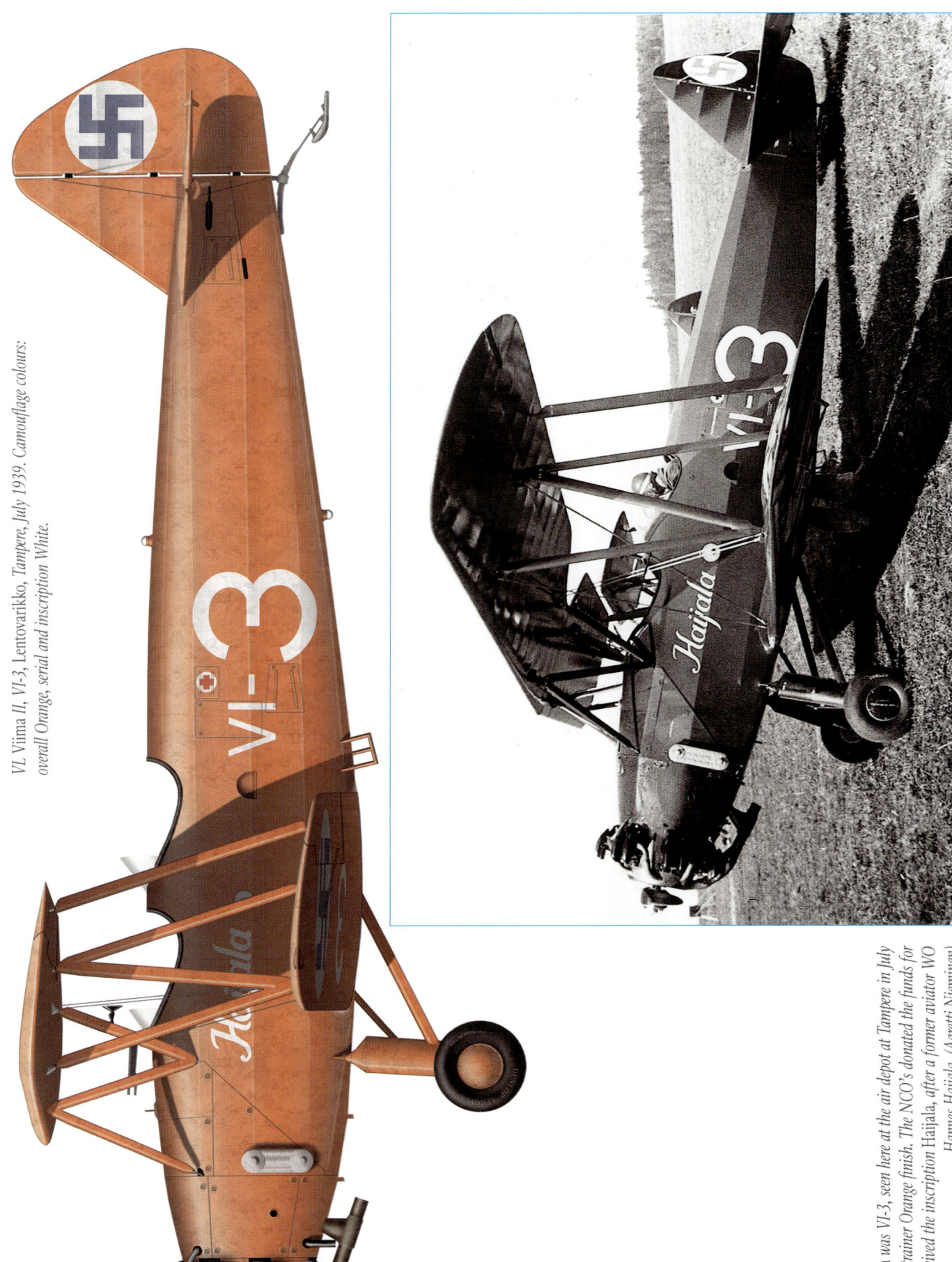

VL Viima II, VI-3, Lentovarikko, Tampere, July 1939. Camouflage colours: overall Orange, serial and inscription White.

The first series Viima was VI-3, seen here at the air depot at Tampere in July 1939, in the overall trainer Orange finish. The NCO's donated the funds for this plane which received the inscription Haijala, after a former aviator WO Hannes Haijala. (Aaretti Nieminen)

VL Viima II, VI-7, Ilmasotakoulu, Kauhava, July 1940. Camouflage colours: overall Orange with spray-on Olive Green, serial White.

Three Viimas, VI-5, 7 and 12 from ISK outside Veljekset Karhumäki factory at Kuorevesi on 12 July 1940, going to major overhaul. The streaky appearance was caused by spraying Olive Green paint on the overall orange colour, to comply with the 12 October 1939 order for camouflage. (Finnish Air Force)

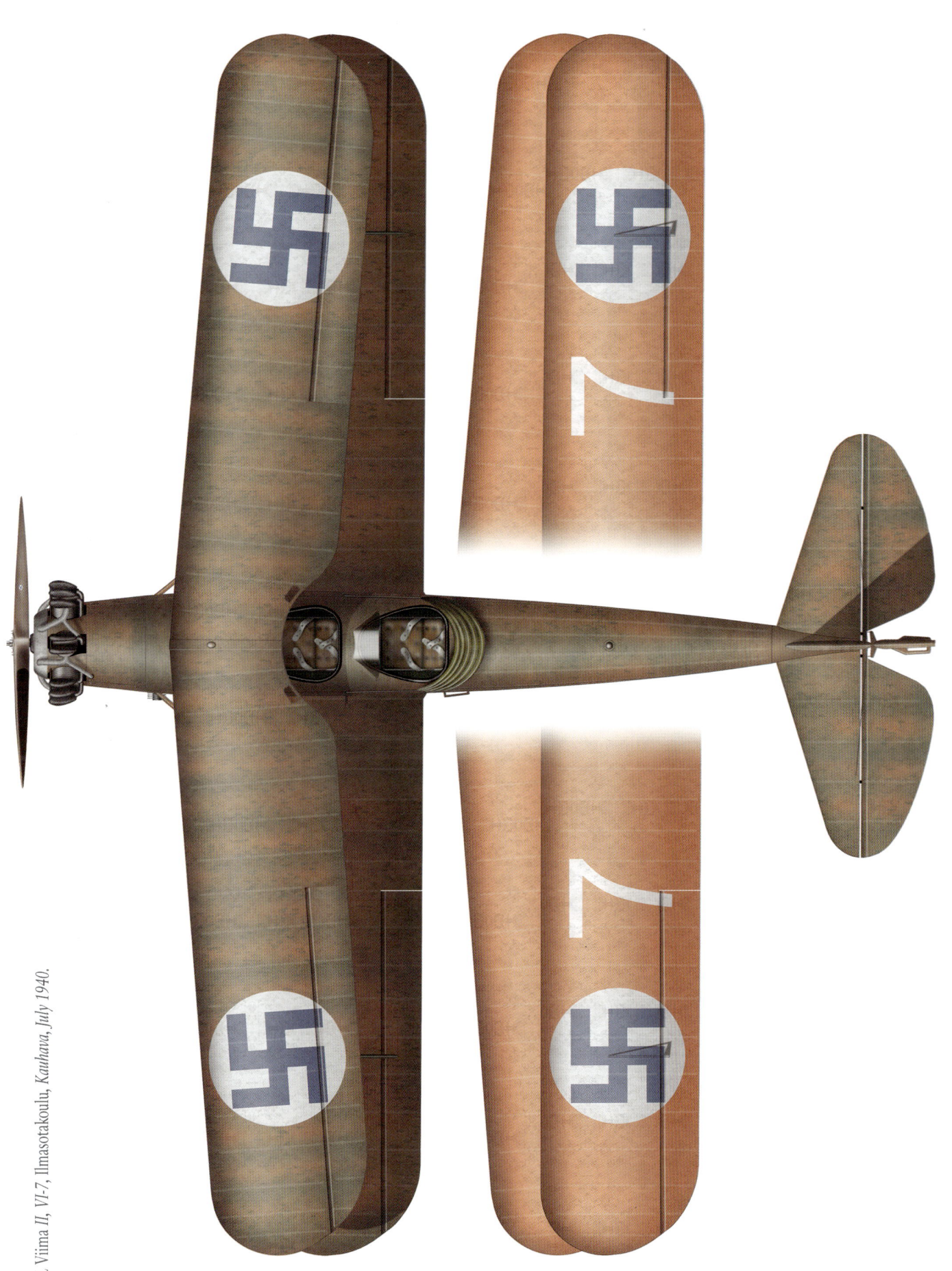

VL Viima II, VI-7, Ilmasotakoulu, Kauhava, July 1940.

VL Viima II, VI-20, Ilmasotakoulu, Kauhava, September 1940. Camouflage colours: Olive Green upper surfaces and aluminium dope under surfaces, serial White.

Viima VI-20 of ISK crashed at Kauhava on 20 September 1940. This machine shows clearly the 12 October 1939 trainer camouflage of Olive Green and aluminium dope. The plane was rebuilt using similarly painted components. (Finnish Air Force)

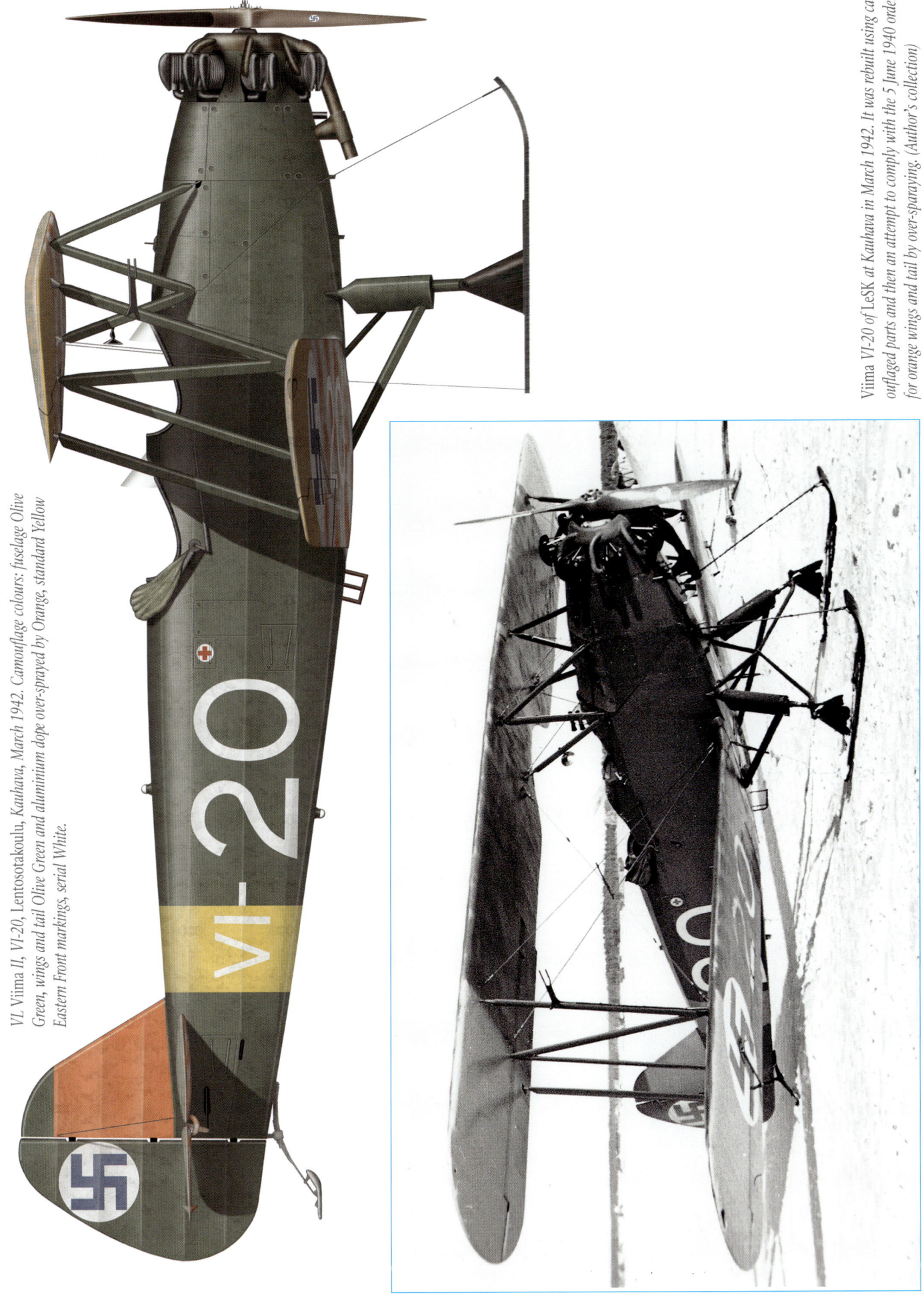

VL Viima II, VI-20, Lentosotakoulu, Kauhava, March 1942. Camouflage colours: fuselage Olive Green, wings and tail Olive Green and aluminium dope over-sprayed by Orange, standard Yellow Eastern Front markings, serial White.

Viima VI-20 of LeSK at Kauhava in March 1942. It was rebuilt using cam-ouflaged parts and then an attempt to comply with the 5 June 1940 order for orange wings and tail by over-spraying. (Author's collection)

VL Viima II, VI-23, Lentosotakoulu, Kauhava, April 1942. Camouflage colours: Olive Green, wings and vertical fin over-sprayed by Orange, standard Eastern Front markings Yellow, serial White.

Viima VI-23 of LeSK after taxiing in a hole at Lappajärvi on 26 April 1942. This was an ex-civil Viima, OH-ILN. When re-painted by Karhumäki the Green on the wings and tail had a smudgy Orange appearance while the serial and fuselage were painted properly; the latter Olive Green. (Author's collection)

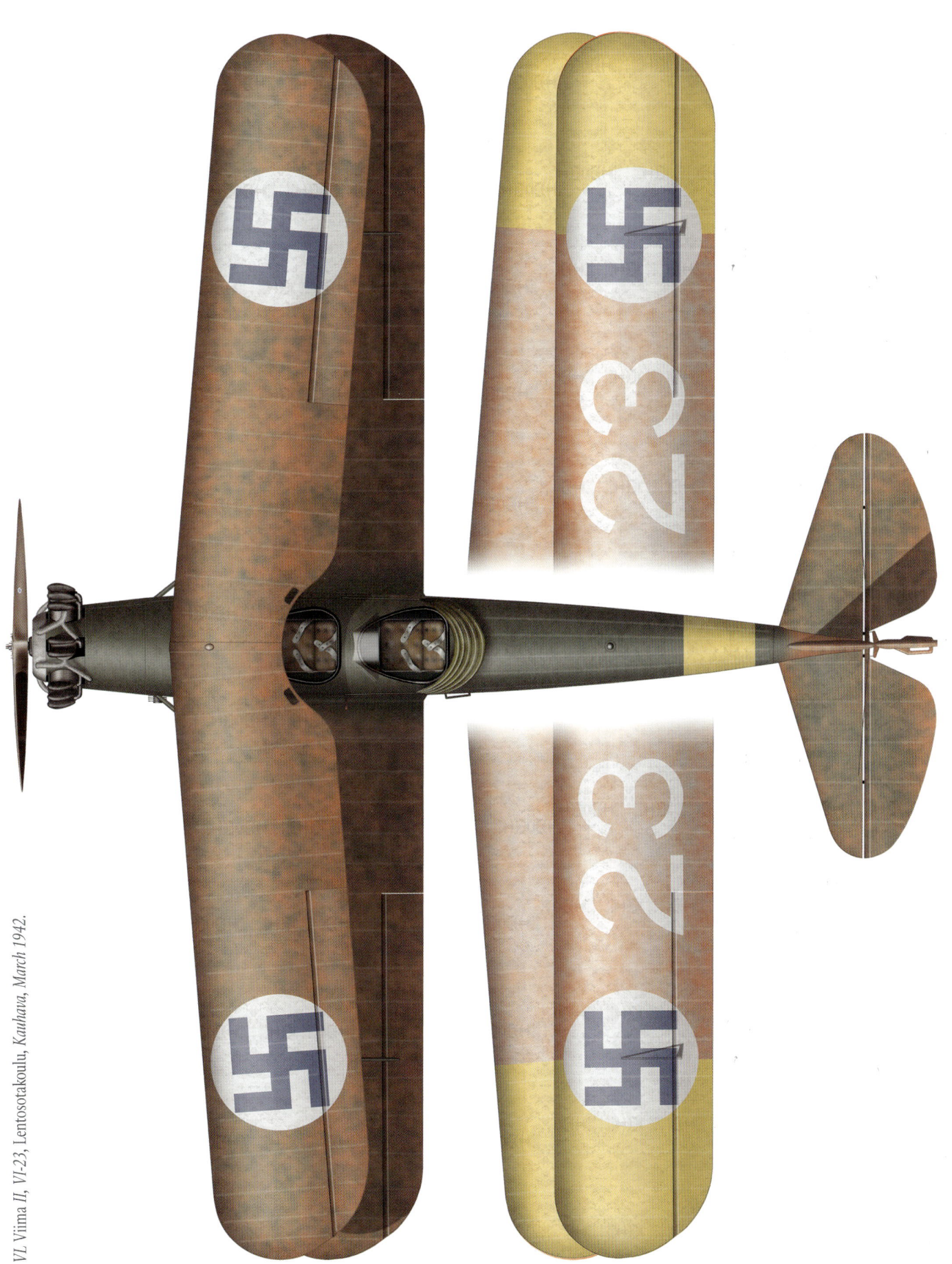

VL Viima II, VI-23, Lentosotakoulu, Kauhava, March 1942.

VL Viima II, VI-4, Lentosotakoulu, Kauhava, June 1942. Camouflage colours: fuselage Olive Green, wings and empennage Orange, standard Eastern Front markings Yellow, serial and inscription White.

Viima VI-4 of LeSK at Kauhava in June 1942 in Olive Green and Orange colours specified for the Viima. The inscription Kokko came from a well-known scholar Väinö Kokko after the elementary school teachers provided the funds for this plane. (Author's collection)

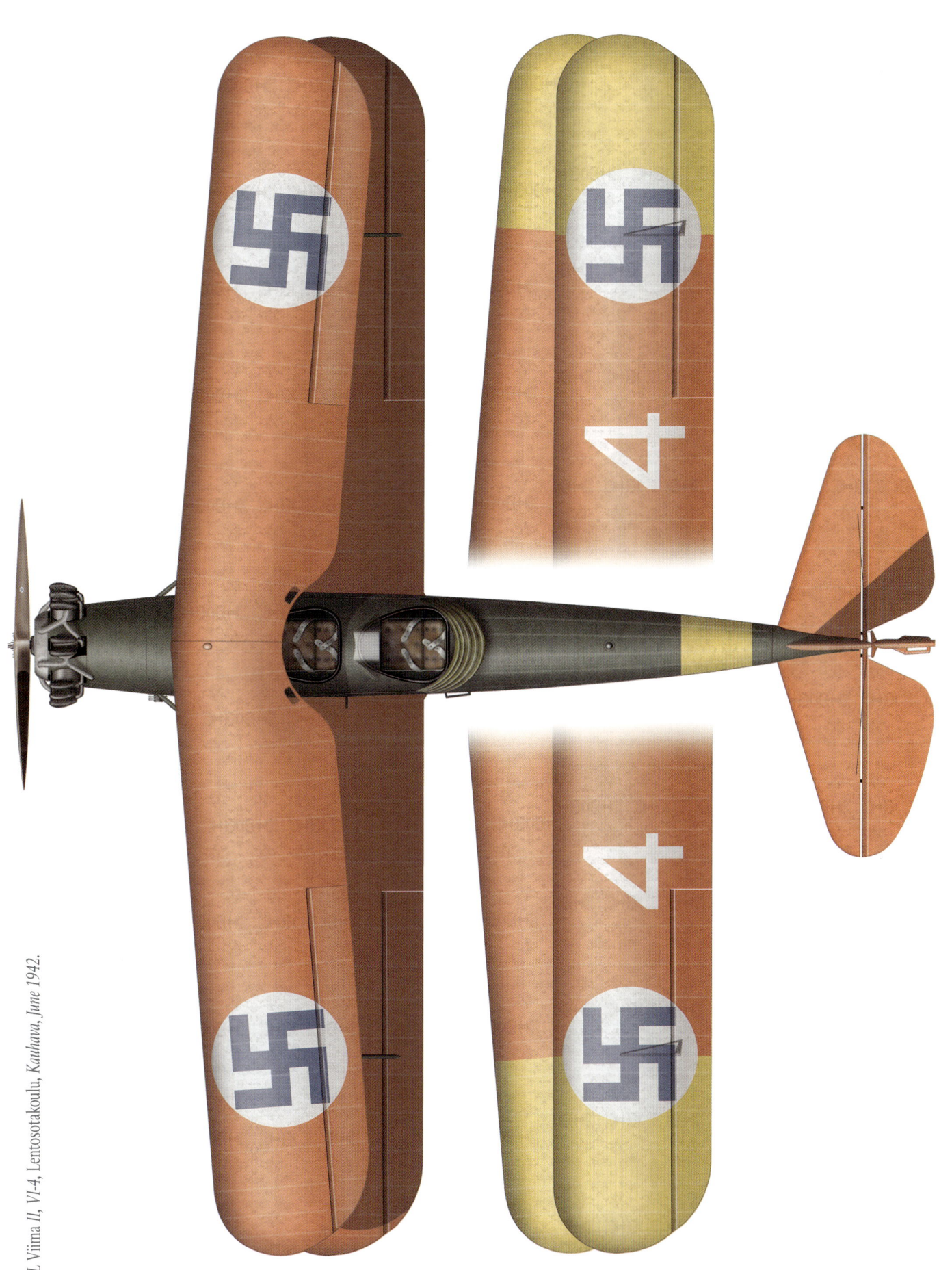

VL Viima II, VI-4, Lentosotakoulu, Kauhava, June 1942.

VL Viima VI-13, Lentolaivue 12, Nurmoila, August 1943. Camouflage colours: Olive Green and Black upper surfaces, DN-colour under surfaces, standard Eastern Front markings Yellow, serial Black.

LeLv 12 had three camouflaged Viimas. Here VI-13 taxis at Nurmoila in August 1943. It has the full Warpaint of Olive Green and Black upper sides with light Blue-Grey DN-colour undersides and regulation Yellow Eastern Front markings. (Lauri Kippo)

VL Viima VI-13, Lentolaivue 12, Nurmoila, August 1943.

Focke-Wulf Fw 44 J *Stieglitz*

Purchase

The German Focke-Wulf Fw 44 *Stieglitz* primary trainer biplane was a design from 1932. It was built in several versions, total production exceeding 1,500. Assisted by the Swedes, thirty aircraft of the export version Fw 44 J were bought from Germany on 12 April 1940. These were serialled SZ-1–SZ-30.

They were flown to Finland from Warnemünde in three ten-aircraft lots. The first batch arrived on 25 May 1940, the next on 7 August 1940 and the last on 14 August 1940. *Českomoravska-Kolben-Danek* in Prague had built the aircraft. Ten of these aircraft (the second lot) were sold on 29 March 1940 to Norway, but never delivered due to the German invasion.

On 7 March 1944 ten additional aircraft were bought from Germany, but only half were delivered, in August 1944, with serials SZ-31–SZ-35. In total the Finnish Air Force had thirty-five Fw 44 J *Stieglitze* trainers in its inventory.

Stieglitz SZ-1 of ISK on arrival at Kauhava on 26 May 1940. These planes were bought jointly with the Swedish Air Force and carried the latter's colours. The fuselage was Dark Grey (Swedish Pansargrå) and the flying surfaces Orange. The serials were also non-standard. (Finnish Air Force)

Stieglitz SZ-9 liaison of the air force HQ parked on the ramp at Helsinki Malmi on 26 June 1940. In August 1941 this plane became the first one to receive Finnish colours, Olive Green fuselage and orange wings and tail plus a large serial. (Finnish Air Force)

Employment

In 1940 twenty of the *Stieglitzes* were placed with the Aviation Fighting School (*LeSK*) as secondary trainers, while the remainder was put into storage until summer 1944. During the Continuation War the *Stieglitzes* served, in addition to *LeSK*, with *T-LLv* 35, an advanced training squadron, while the air force HQ flew four aircraft and the Test Flight one.

After the war 30 *Stieglitzes* remained on duty in 1945, now as a primary trainer at the Air Fighting School. Additionally they were used as liaison aircraft in all other air force units. Three aircraft survived into the 1960s and SZ-25 flew the last flight of the type on 30 August 1960. 20 *Stieglitzes* were sold at auction and fifteen aircraft were later registered to air clubs.

Colours

On arrival the Focke-Wulf Fw 44 *Stieglitzes* had dark grey fuselages and orange flying surfaces, according to the Swedish practise.

On 5 June 1940 new instructions were issued. The fuselage of the *Stieglitzes* was to be painted Olive Green while the wings and tail were to be orange. Three *Stieglitzes*, SZ-9, 5 and 7, received these colours

SZ-15 of LeSK at Kauhava in March 1942. Only the regulation Eastern Front markings were added, yellow fuselage band and undersides of both wing tips. Under the stabilizer is a data block. (Author's collection)

SZ-30 was the courier plane of the air force HQ and is seen here on a visit to the air depot at Kuorevesi in summer 1942. It still wears the original factory finish of Dark Grey and Orange. (Author's collection)

in the latter half of 1941. In 1942 SZ-13 was also repainted and in 1943 SZ-10. In 1944 this scheme was painted in turn on SZ-11, 24, 4, 23, 8, 30 and 26. After the war the remaining nineteen were painted this way. This scheme was carried as long as the *Stieglitzes* flew.

An exception was the last batch of five planes arriving in August 1944 as some of them were camouflaged in RLM colours 70/71/65.

Serials

On arrival the factory serials were large and in white, not according to the 20 March 1934 regulation. On 13 May 1939 small letters and large numbers in white were ordered to be painted on basic trainers. The numbers were also introduced under both lower wings. This also transferred the national marking from the fuselage to the rudder. This practise was extended to cover *Stieglitzes*, when they went for a major overhaul or repair and was to remain throughout the careers of the type.

S/n	C/n	Delivered	Struck off charge	Remarks	Hours
SZ-1	2880	26 May 1940	31 Aug 1942	W/o Kauhava 10 Jul 1942	571.05
SZ-2	2891	26 May 1940	31 May 1960	Into storage 19 Feb 1958	3926.40
SZ-3	2894	26 May 1940	14 Jan 1956	W/o Mustasaari 14 Mat 1955	2758.10
SZ-4	2895	26 May 1940	31 May 1960	Into storage 10 Oct 1958	3212.30
SZ-5	2896	26 May 1940	31 May 1960	Into storage 27 May 1958	2879.10
SZ-6	2897	26 May 1940	6 Nov 1941	W/o Kauhava 26 Sep 1941	463.30
SZ-7	2903	26 May 1940	18 Dec 1958	W/o Kauhava 19 Sep 1958	3325.30
SZ-8	2905	26 May 1940	18 Dec 1959	W/o Luumäki 19 Sep 1957	2397.50
SZ-9	2906	1 Jun 1940	31 May 1960	Into storage 27 Jan 1959	3833.40
SZ-10	2914	26 May 1940	31 May 1960	Into storage 20 Mar 1958	3228.20
SZ-11	2780	8 Aug 1940	31 May 1960	Into storage 6 Apr 1956	2884.05
SZ-12	2775	8 Aug 1940	31 May 1960	Into storage 5 Sep 1959	4193.20
SZ-13	2781	8 Aug 1940	26 Apr 1944	W/o Kauhava 3 Apr 1944	156.45
SZ-14	2776	8 Aug 1940	31 May 1960	W/o Luonetjärvi 16 Feb 1959	3399.10
SZ-15	2782	8 Aug 1940	31 May 1960	Into storage 14 Jul 1959	3548.45
SZ-16	2777	8 Aug 1940	17 Oct 1944	W/o Vesivehmaa 9 Nov 1943	669.15
SZ-17	2826	8 Aug 1940	17 Oct 1944	W/o Kauhava 1 Nov 1942	541.35
SZ-18	2778	8 Aug 1940	31 May 1960	Into storage 10 Mar 1959	3543.05
SZ-19	2827	8 Aug 1940	31 May 1940	Into storage 30 Dec 1959	3473.50
SZ-20	2779	8 Aug 1940	31 May 1960	Into storage 6 Nov 1958	3233.35
SZ-21	2924	11 Nov 1943	31 May 1960	Into storage 15 Oct 1958	3337.20
SZ-22	2925	22 May 1944	16 Feb 1954	W/o Kauhava 18 Jan 1954	2029
SZ-23	2929	7 May 1943	19 Oct 1959	W/o Helsinki 10 Jul 1959	3069.35
SZ-24	2927	16 Aug 1940	31 May 1960	Into storage 12 Jan 1959	3677.30
SZ-25	2928	10 Sep 1943	3 Nov 1960	Last flight 30 Aug 1960	3434.20
SZ-26	2926	9 Jun 1941	31 May 1960	Into storage 13 May 1950	3306.05
SZ-27	2930	6 Sep 1942	18 Dec 1958	W/o Kauhava 30 Oct 1956	2387.50
SZ-28	2931	22 May 1944	18 Dec 1958	W/o Kauhava 19 Sep 1958	2514.20
SZ-29	2932	5 Dec 1940	31 May 1960	Into storage 9 Jun 1958	3108.10
SZ-30	2933	29 Jan 1942	31 May 1960	Into storage 14 May 1960	3432.45
SZ-31	SD+QC	3 Aug 1944	31 May 1960	Into storage 28 Jan 1959	3038.50
SZ-32	SD+QZ	3 Aug 1944	31 May 1960	Into storage 6 Oct 1959	3267.35
SZ-33	BB+EL	3 Aug 1944	31 May 1960	Into storage 30 Sep 1958	3367.10
SZ-24	TY+BW	14 Aug 1944	18 Dec 1958	W/o Kauhava 9 Sep 1958	3282
SZ-35	BB+EJ	14 Aug 1944	31 May 1960	Into storage 10 Oct 1958	3172.40

Stieglitz SZ-11 at the air depot at Kuorevesi in March 1944. It was repaired after a landing accident by Karhumäki and received the regulation paintwork of Olive Green fuselage and Orange wings and tail. On 25 May 1944 it was delivered to LeSK. (Finnish Air Force)

Focke Wulf Fw 44J Steglitz, SZ-3, Ilmasotakoulu, Kauhava, May 1940. Camouflage colours: fuselage colour Pansargrå (Dark Grey), wings and horizontal stabilizer Orange.

SZ-3 of ISK on the ramp mat Kauhava on 27 May 1940. It has the Swedish style trainer colours of Dark Grey fuselage and Orange flying surfaces. Even the wheel caps are Orange. (Finnish Air Force)

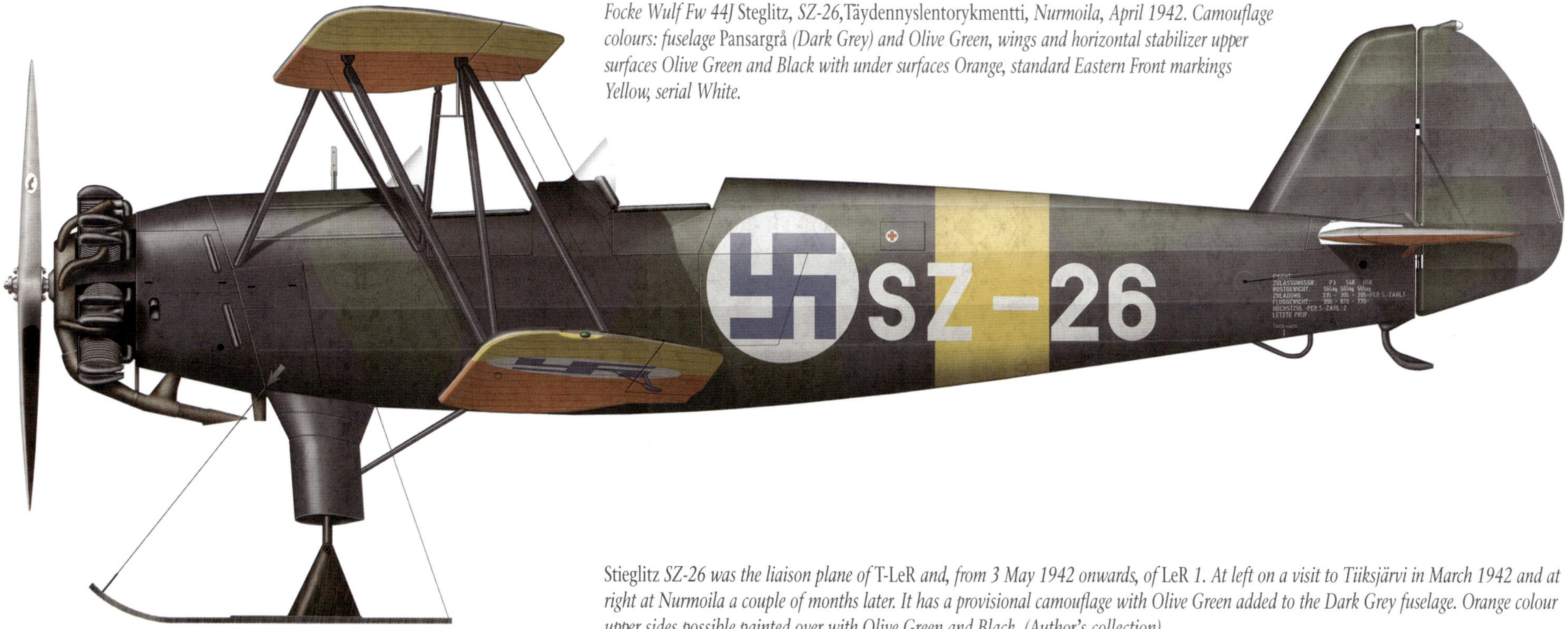

Focke Wulf Fw 44J Steglitz, SZ-26, Täydennyslentorykmentti, Nurmoila, April 1942. Camouflage colours: fuselage Pansargrå (Dark Grey) and Olive Green, wings and horizontal stabilizer upper surfaces Olive Green and Black with under surfaces Orange, standard Eastern Front markings Yellow, serial White.

Stieglitz SZ-26 was the liaison plane of T-LeR and, from 3 May 1942 onwards, of LeR 1. At left on a visit to Tiiksjärvi in March 1942 and at right at Nurmoila a couple of months later. It has a provisional camouflage with Olive Green added to the Dark Grey fuselage. Orange colour upper sides possible painted over with Olive Green and Black. (Author's collection)

Focke Wulf Fw 44 J Steglitz, SZ-2, Lentosotakoulu, Kauhava, April 1942. Camouflage colours: fuselage Pansargrå (Dark Grey), wings and horizontal stabilizer Orange, standard Eastern Front markings Yellow, serial White.

(Left) SZ-2 of LeSK still on skis at Kauhava on 12 May 1942. It has the original Swedish style Dark Grey and Orange colours. (Right) SZ-2 of T-LeLv 35 after a bad night landing at Vesivehmaa on 24 September 1943. Only the yellow Eastern Front markings have been added. (Finnish Air Force)

Focke Wulf Fw 44 J Steglitz, SZ-2, Lentosotakoulu, Kauhava, April 1942.

Focke Wulf Fw 44J Stieglitz, SZ-9, Lentosotakoulu, Kauhava, July 1943. Camouflage colours: fuselage Olive Green, wings, horizontal stabilizer and tail Orange, standard Eastern Front markings Yellow, serial White.

SZ-9 of LeSK at Kauhava in July 1943. It has the 5 June 1940 order colours of Olive Green fuselage and Orange wings and tail with the large serial specified on 13 May 1939. This was the first Stieglitz to get this paintwork, in August 1941. (SA-kuva))

Focke Wulf Fw 44J Stieglitz, SZ-9, Lentosotakoulu, Kauhava, July 1943.

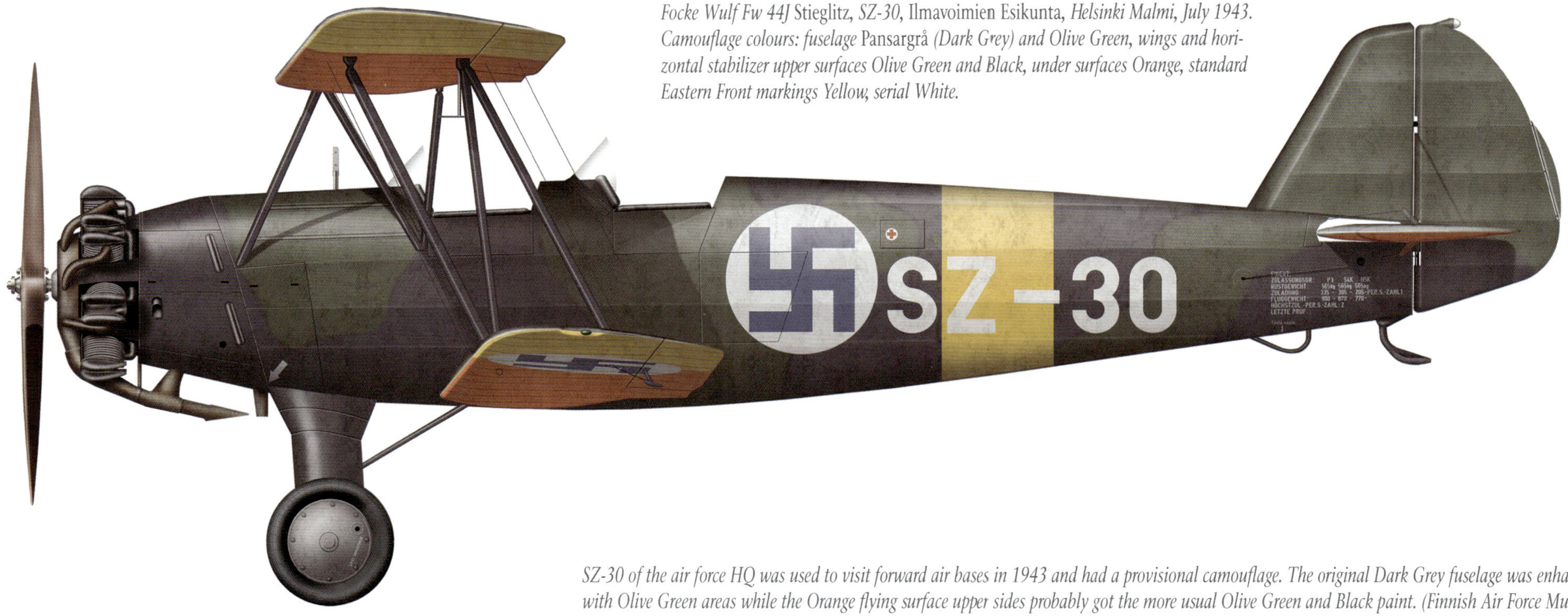

Focke Wulf Fw 44J Stieglitz, SZ-30, Ilmavoimien Esikunta, Helsinki Malmi, July 1943. Camouflage colours: fuselage Pansargrå (Dark Grey) and Olive Green, wings and horizontal stabilizer upper surfaces Olive Green and Black, under surfaces Orange, standard Eastern Front markings Yellow, serial White.

SZ-30 of the air force HQ was used to visit forward air bases in 1943 and had a provisional camouflage. The original Dark Grey fuselage was enhanced with Olive Green areas while the Orange flying surface upper sides probably got the more usual Olive Green and Black paint. (Finnish Air Force Museum)

Focke Wulf Fw 44J Stieglitz, SZ-30, Ilmavoimien Esikunta, Helsinki Malmi, July 1943.

Stieglitz SZ-33 was the liaison of HLeLv 33. It is seen here at Utti on 24 February 1950, after a landing with a malfunctioning ski. The markings comply fully with the 13 May 1939 order of larger White serials and the 5 June 1940 order of an Olive Green fuselage and Orange wings and tail. Only the national insignias are of the 1 April 1945 type roundels. (Author's collection)

Advanced Trainers

- Gloster Gamecock I and II
- ASJA *Jaktfalk* II
- Bristol Bulldog II and IVA
- Gloster Gauntlet II
- Polikarpov I-15bis

- VL *Pyry* I
- Fokker D.XXI
- Fokker C.X
- FIAT G.50

Gloster Gamecock I and II

Purchase

The Gamecock was a British fighter design from 1925. It was built in two versions, with the total production reaching 94 aircraft. Finland became interested in the Gamecock and in March 1927 obtained one trial Gamecock I, serial GA-38. A manufacturing licence was acquired in April 1928 and one Gamecock II pattern aircraft, GA-43, was bought in December 1929. The State Aircraft Factory built fifteen Gamecock II fighters by May 1930, which were serialled GA-44–GA-58.

Thus the Finnish Air Force had a total of seventeen Gamecocks in its inventory.

The first Gamecock obtained was GA-38, with the short fuselage of Mark I and new wings of Mk II. It is here in front of MLE hangar at Utti during summer 1927. All markings are by the aircraft factory standards. (Finnish Air Force)

Gamecock GA-45 of MLE at Utti in spring 1930. The overall factory finish is aluminium dope with the nose and spine in NIVO, an olive green, which later became the standard Finnish camouflage colour. (Finnish Air Force)

Employment

The new Gamecocks were positioned at *Maalentoeskaaderi* (Land Escadre), which became in July 1933 *Lentoasema* 1 (Air Station). From October 1934 onwards the Gamecocks served with Lentolaivue 24, the premium fighter squadron.

When *LLv* 24 was equipped with the Dutch Fokker D.XXIs from December 1937 onwards, the Gamecocks were posted to *Ilmasotakoulu* (Air Fighting School) from September 1938 onwards, a total of nine fighters. When the Winter War broke out on 30 November 1939, the remaining five Gamecocks became advanced trainers with *T-LLv* 29, which was part of *T-Lento*R 2. This again was the training element of the fighter regiment LentoR 2.

After the Winter War the temporary T-Lento R 2 was disbanded on 27 March 1940 and the Gamecocks posted to *T-LLv* 35, which was a new advanced training squadron. At the beginning of the Continuation War in June 1941 the four remaining Gamecocks were delivered back to *Lentosotakoulu* (ex-*ISK*). Two were damaged beyond repair in the second half of 1941, the third was put into storage in 1943 and the flying career of the Gamecock ended in a bad landing of GA-46 on 22 July 1944.

Gamecock GA-43 of LLv 24 camouflaged under the nets at Utti on 1 November 1935. The plane has an experimental splinter pattern camouflage of Green and Reddish Brown. (Finnish Air Force)

A Gamecock line-up of LLv 24 in an aviation parade held at Suur-Merijoki on 3 August 1935. All planes have gone via the aircraft factory and received the standard camouflage of Olive Green tops and aluminium dope undersides. (Finnish Air Force)

Colours

The factory finish for the Gamecock was an overall aluminium dope with NIVO (Olive Green) nose and upper decking.

The pre-war standard camouflage of olive green upper sides and aluminium dope lower sides was introduced with the *Tuisku* trainer in January 1934. When the Gamecocks entered service with ISK by early 1939, as advanced trainers, all remaining planes carried this scheme but with light grey undersides, up to the end of their flying career.

Serials

The Gamecocks wore the standard factory serials until 20 March 1934, when a new directive stipulated that the height of the serial was to be 2½ times of the arm of the fuselage swastika, much smaller than before. All serials were black in colour and the style according to the Finnish standard SFS Z.I.1. These serials were to remain on the Gamecocks up to their write-off.

The landing of GA-46 of ISK ended on its nose at Kauhava on 16 December 1938. This view clearly shows the position and size of the wing insignia, as well as the solid Olive Green camouflage. (Author's collection)

GA-55 of ISK flipped over in a landing to Kauhava on 23 May 1939. The aircraft displays the camouflage and markings, which were applied fully by the book. (Author's collection)

S/n	C/n	Delivered	Struck off charge	Remarks	Hours
GA-38		17 Aug 1939	11 Oct 1941	W/o Siikakangas 6 Aug 1941	
GA-43		11 Oct 1939	11 Oct 1941	W/o Parola 20 Dec 1939	
GA-45	2	28 Dec 1938	9 Apr 1940	W/o Parola 24 Feb 1940	
GA-46	3	26 Sep 1938	30 Aug 1944	W/o Vaasa 22 Jul 1944	
GA-49	6	18 Oct 1939	30 Dec 1939	W/o Parola 11 Dec 1939	
GA-50	7	4 Mar 1939	20 Dec 1942	W/o Kauhava 21 Oct 1942	
GA-51	8	26 Sep 1938	31 Oct 1939	W/o Laajalahti 7 Sep 1939	
GA-55	12	14 Feb 1939	27 Sep 1941	W/o Siikakangas 8 Aug 1941	
GA-58	15	12 Nov 1939	9 Apr 1940	W/o Parola 10 Mar 1940	

This table lists only those Gamecocks which flew in the advanced trraining role, in the given period.

Gamecock GA-58 of ISK poses for the official photographer at Kauhava on 21 November 1938. The aircraft was painted only a few weeks earlier and by this time the factory had begun using the light grey underside colour, for planes with fabric or wooden skinning, instead of the regulation aluminium dope. (Finnish Air Force)

GA-46 was the longest-lived example of its type. It is seen here at Parola with T-LLv 29 in early December 1939. The white circles of the national insignia have been covered with Olive Green paint for better camouflage effect (unofficially), only to be washed clean soon after. (Author's collection)

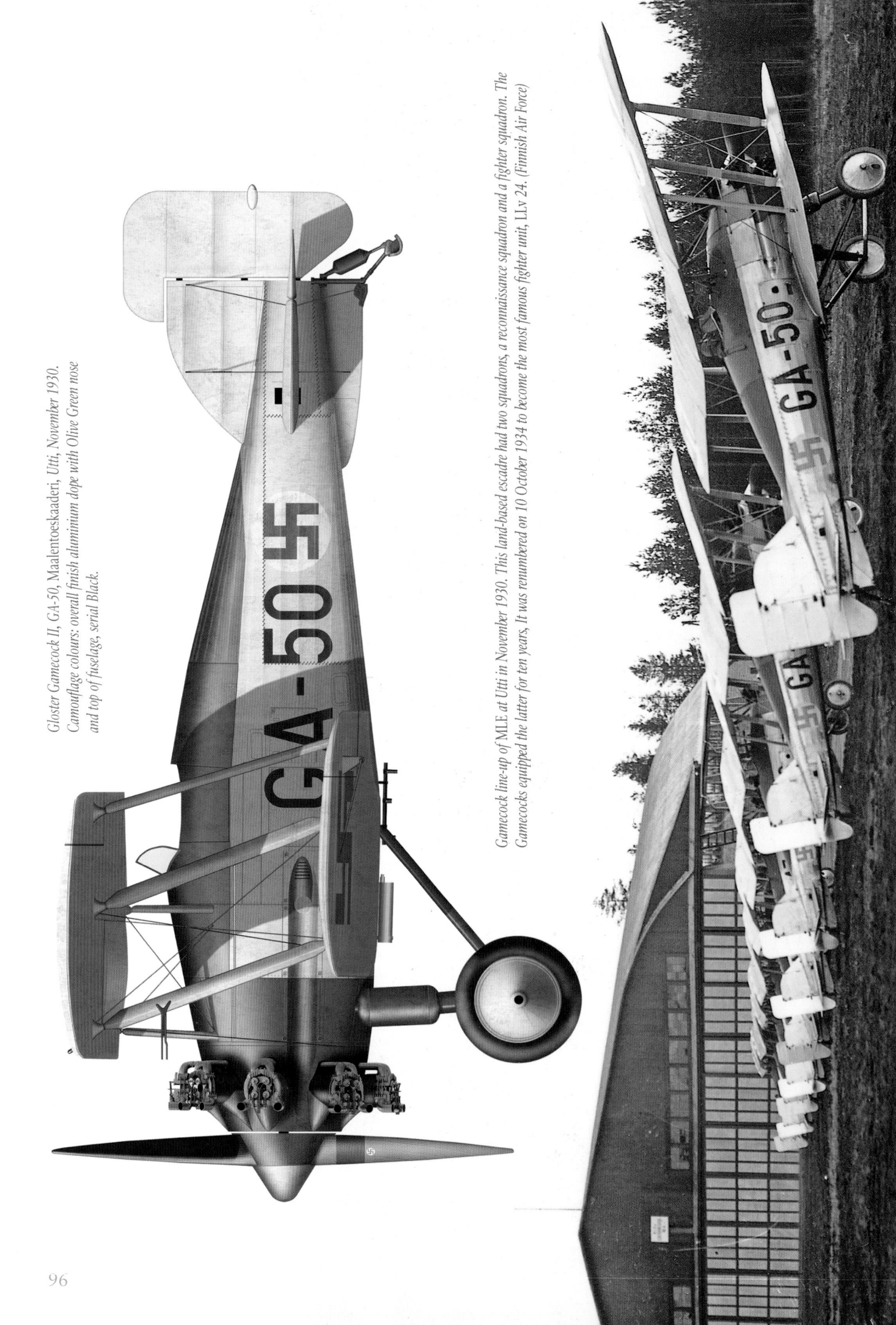

Gloster Gamecock II, GA-50, Maalentoeskaaderi, Utti, November 1930. Camouflage colours: overall finish aluminium dope with Olive Green nose and top of fuselage, serial Black.

Gamecock line-up of MLE at Utti in November 1930. This land-based escadre had two squadrons, a reconnaissance squadron and a fighter squadron. The Gamecocks equipped the latter for ten years, It was renumbered on 10 October 1934 to become the most famous fighter unit, LLv 24. (Finnish Air Force)

Gloster Gamecock II, GA-50, Maalentoeskaaderi, Utti, November 1930.

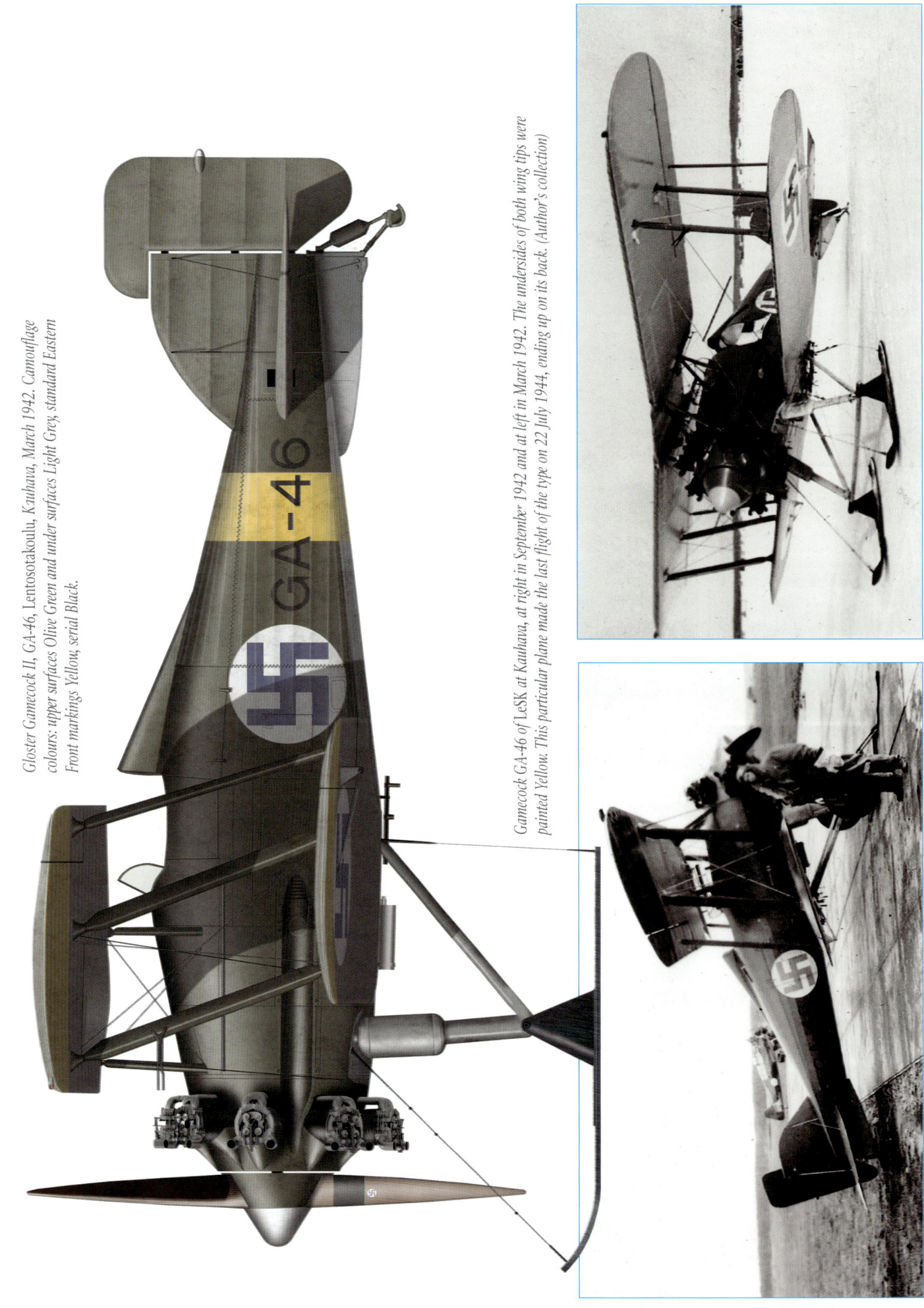

Gloster Gamecock II, GA-46, Lentosotakoulu, Kauhava, March 1942. Camouflage colours: upper surfaces Olive Green and under surfaces Light Grey, standard Eastern Front markings Yellow, serial Black.

Gamecock GA-46 of 1eSK at Kauhava, at right in September 1942 and at left in March 1942. The undersides of both wing tips were painted Yellow. This particular plane made the last flight of the type on 22 July 1944, ending up on its back. (Author's collection)

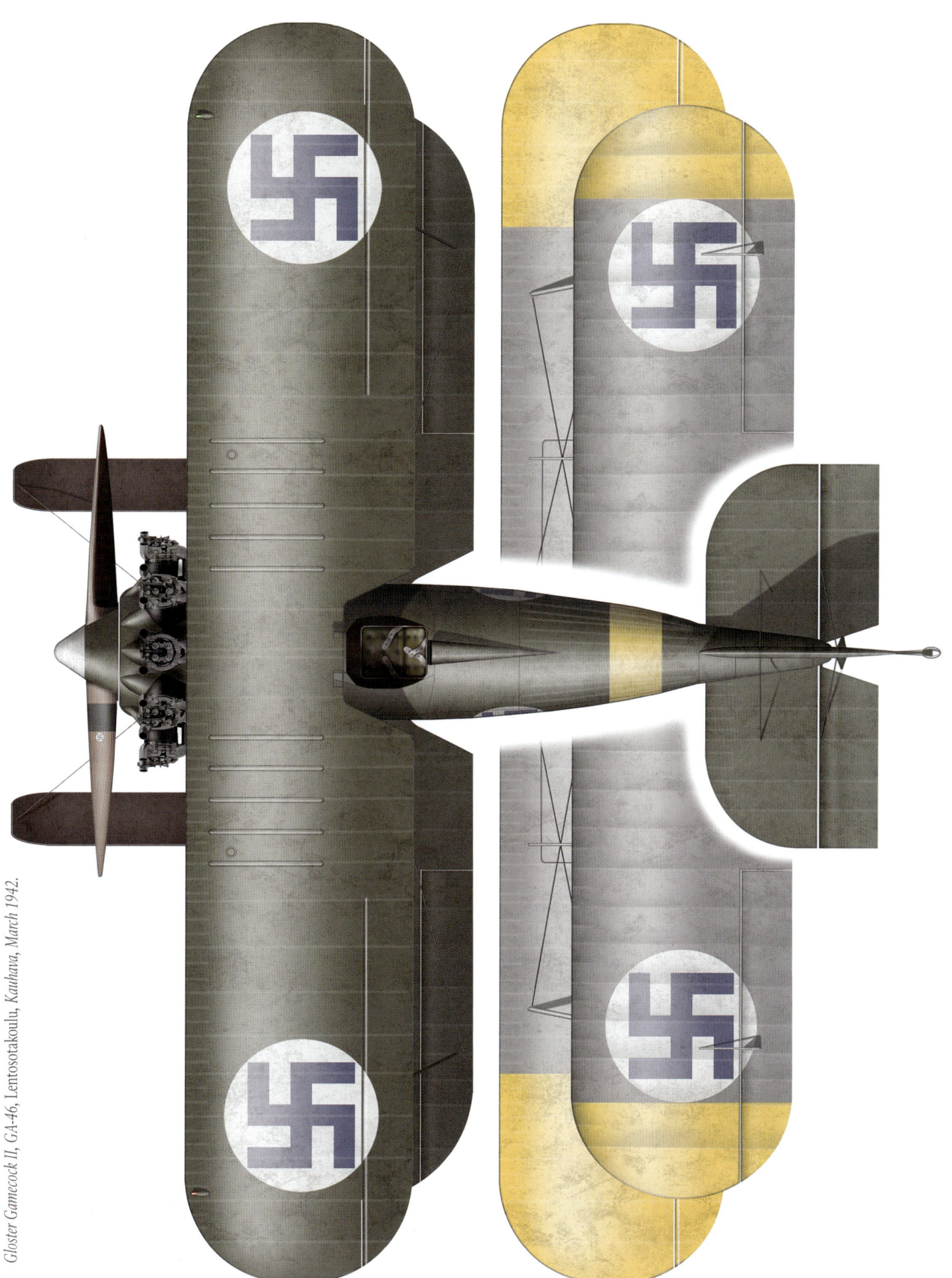

Gloster Gamecock II, GA-46, Lentosotakoulu, Kauhava, March 1942.

ASJA *Jaktfalk* II

Purchase

The *Jaktfalk* was a Swedish fighter design from 1930. Two factories constructed a total of 18 aircraft, Svenska Aero built eleven and ASJA seven. After the beginning of the Winter War Sweden donated three *Jaktfalk* II aircraft, which arrived in Finland on 15 December 1939. They were given Finnish serials after the Swedish ones, JF-219, 224 and 228.

Employment

The three *Jaktfalks* went straight to the advanced training squadron *T-LLv* 29 and their service continued in other similar units and with *LeSK* (Air Fighting School). Due to accidents and lack of spares the planes were grounded for months. JF-219 was damaged beyond repair in August 1941, JF-224 became a write-off in February 1944 and JF-228 ended its career with a flight to storage on 20 February 1945.

Colours

On arrival the *Jaktfalks* were in the factory finish of overall aluminium dope. All three suffered accidents early in their careers and during the repair were painted in the *Tuisku* trainer camouflage of olive green tops and (most likely) light grey undersides. JF-219 was to first the receive this camouflage in February 1941, followed by JF-224 in June 1941 and JF-228 in November 1941, keeping these colours to the end of their flying career.

Jaktfalk II No. 228 of T-LLv 29 at Tyrväntö on March 1940. This plane was built by ASJA and had the Swedish Air Force designation J6B and serial 228. Quite exceptionally only the blue swastikas of the national insignia have been applied. This became JF-228 in Finnish service. (Reino Lampelto)

JF-219 was a Svenska Aero built J6A Jaktfalk II. It is parked here at Siikakangas in early June 1941, belonging to LLv 34. The plane has olive green upper surfaces and light grey lower surfaces with proper serials, applied during damage repair four months earlier. (Heimo Lampi)

JF-224 of T-LLv 35 at Vesivehmaa in June 1942. It has full regulation markings and colours with factory painted Light Grey under surfaces. Yellow Eastern Front markings (consisted of the 50 cm rear fuselage band and only the underside of the lower wing tips). (Olli Riekki)

Serials

On arrival in Finland, the only serial present was the small Swedish number in front of the stabilizer. In the accident repairs the *Jaktfalks* received the standard serials as specified on 20 March 1934, the height of the serial was 2½ times of the arm of the fuselage swastika. The font complied with the Finnish standard SFS Z.I.1.

S/n	C/n	Delivered	Struck off charge	Remarks	Hours
JF-219	219	15 Dec 1939	15 Dec 1942	W/o Pori 7 Aug 1941	65.25
JF-224	224	15 Dec 1939	14 Mar 1944	W/o Lapua 22 Feb 1944	226.20
JF-228	228	15 Dec 1939	3 Jun 1946	Last flight 20 Feb 1945	373.20

This accident of JF-228 hjas remained as a mystery. The aircraft shows the regular camouflage of Olive Green upper and Light Grey lower surfaces, dating the photo before mid-June 1941, the location being possibly Pori (Author's collection)

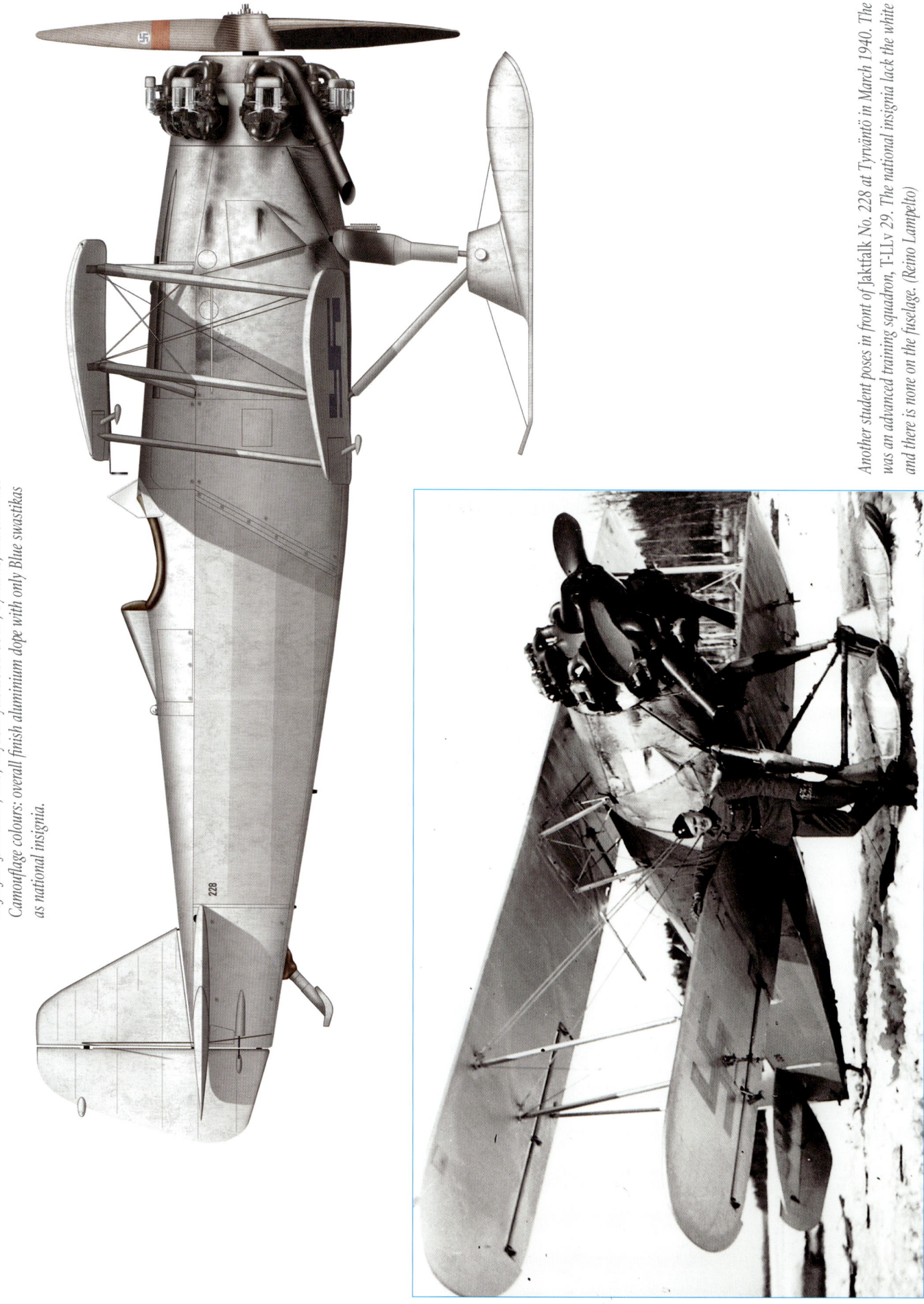

ASJA J6B Jaktfalk II, 228, Täydennyslentolaivue 29, Tyrväntö, March 1940.
Camouflage colours: overall finish aluminium dope with only Blue swastikas
as national insignia.

Another student poses in front of Jaktfalk No. 228 at Tyrväntö in March 1940. The unit
was an advanced training squadron, T-LLv 29. The national insignia lack the white circle
and there is none on the fuselage. (Reino Lampelto)

ASJA J6B Jaktfalk II, JF-224, Lentosotakoulu, Kauhavg, July 1943. Camouflage colours: upper surfaces Olive Green, under surfaces Light Grey, standard Eastern Front markings Yellow, serial Black.

Jaktfalk JF-224 in different seasons. At left with LeSK at Kauhava on 19 February 1943 and at right also with LeSK at Kauhava on 10 July 1943. It wears the regulation colours and markings. (Finnish Air Force and SA-kuva)

Bristol Bulldog II and IVA

Purchase

The Bulldog was a British fighter design from 1927. It was built in two major version in a total of 443 examples. Finland purchased in March 1934 seventeen Bulldog IVA aircraft for the recently established second pure land-based fighter squadron. These planes were delivered the following February with Finnish Air Force serials BU-59–BU-75 inclusive.

After the outbreak of the Winter War Sweden donated two Bulldog II aircraft, both arriving in Finland on 15 December 1939. BUj-214 crashed on 2 March 1940 and BUj-216 served until a crash on 9 November 1942.

The first Finnish Bulldog IVA, serial BU-39, ready for delivery at Filton, England on 30 December 1934. The camouflage was the regulation Olive Green upper surfaces and aluminium dope lower surfaces. (Bristol)

Brand new Bulldog BU-62 at the air depot at Santahamina outside Helsinki on 20 February 1935. Exactly a month later it was handed over to fighter squadron LLv 26. It wears the standard Olive Green and aluminium dope camouflage specified in January 1934. (Finnish Air Force)

Employment

In spring 1935 the Bulldogs were assigned to *Lentolaivue 26*, the fighter element of of Air Station 5 based at Suur-Merijoki next to Viipuri. On 1 January 1938 the air stations became aviation regiments and in this case *LLv 26* was transferred to *LentoR 2*, where they served when the Winter War began on 30 November 1939.

On 2 February 1940 the remaining five Bulldogs were handed over to *T-LLv 29*, which was the training unit of its regiment. After the Winter War the Bulldogs continued to serve in the training role with *T-LLv 35* and *LLv 34*, both part of the new aviation regiment, *LentoR 3*.

At the outbreak of the Continuation war on 25 June 1941 these training squadrons possessed eight Bulldogs. Some of them served later with *T-LLv 17*, before all were posted to *LeSK* (Aviation Fighting School) in June 1942. Three Bulldogs crashed that year, two in the next year and the last flight was performed by BU-59 on 22 February 1944.

A line-up of ten LLv 26 Bulldogs at an aviation parade held at the unit's home base Suur-Merijoki on 3 August 1935. Apart from the serial numbers, individual planes could be distinguished by different colours on the propeller spinner. (Finnish Air Force)

Bulldog BU-65 of LLv 26 in a rare pose at Utti on 5 March 1938. While taxiing for take-off a sudden breeze caught the plane and overturned it. The damage was minimal. (Finnish Air Force)

BU-73 of T-LLv 17 on a taxi strip at Pori shortly before the delivery to T-LLv 35, which took place on 3 October 1941. Apart from the fuselage band, only the lower wing tip underside was painted Yellow. (Kullervo Virtanen)

Colours

The Bulldogs were manufactured in Great Britain and painted according to the recently issued Finnish camouflage specifications: upper sides in Olive Green and lower sides in aluminium dope.

During a flying accident repair or major overhaul, the State Aircraft Factory usually replaced all fabric of the aircraft, and painted the top sides with Olive Green and lower sides Light Grey. The first Bulldog to be so painted was BU-65 in October 1938. Two more, BU-63 and 71 followed before the Winter War and by the Continuation War a total of fourteen Bulldogs had been re-painted.

Only one Bulldog, BU-65, received the Warpaint: Olive Green upper side with broad Black stripes and Light Grey lower side, which was painted on 10 July 1942. The Bulldog was by then specified as a trainer, which kept the Light Grey under surfaces.

Serials

The Bulldogs wore the standard serials as specified on 20 March 1934, the height of the serial was 2½ times of the arm of the fuselage swastika. The font complied with the Finnish standard SFS Z.I.1.

Bulldog BU-63 of T-LLv 35 at Vesivehmaa on March 1942. On 17 April 1942 it was flown to the air depot, being totally worn-out and ready for write-off. The colours and markings comply with all regulations. (Eino Ritaranta collection)

BU-216 was one of two Bulldog IIs donated by Sweden in December 1939. It is seen here with T-LeLv 35 at Vesivehmaa in May 1942. As with most non-combat aircraft, the camouflage applied by the State Aircraft Factory was Olive Green upper and Light Grey lower sides. (Olli Riekki)

S/n	C/n	Delivered	Struck off charge	Remarks	Hours
BU-59		12 Apr 1940	11 May 1944	W/o Kauhava 22 Feb 1944	
BU-61		2 Feb 1940	13 Apr 1940	W/o Parola 2 Mar 1940	
BU-62		14 Apr 1940	1 Aug 1941	W/o Parola 25 Jun 1941	
BU-63		2 Feb 1940	16 Aug 1942	Last flight 17 Apr 1942	
BU-65		14 Sep 1940	16 Feb 1944	W/o Kauhava 26 Jan 1944	
BU-66		7 Feb 1940	2 Nov 1940	W/o Parola 25 Mar 1940	
BU-67		31 Jan 1940	20 Dec 1942	W/o Kauhava 28 Oct 1942	
BU-68		2 Feb 1940	16 Aug 1943	W/o Kauhava 29 Jun 1943	
BU-70		16 Mar 1940	27 Jul 1941	W/o Tyrväntö 17 Apr 1940	
BU-71		25 May 1940	20 Dec 1942	W/o Kauhava 13 Oct 1942	
BU-72		2 Feb 1940	10 May 1940	W/o Tyrväntö 23 Mar 1940	
BU-73		2 Feb 1940	30 Jan 1942	W/o Asikkala 18 Dec 1941	
BU-74		2 Feb 1940	19 Sep 1942	W/o Kokkola 17 Jul 1942	
BU-75		11 May 1940	16 Sep 1943	Into storage 9 Dec 1941	
BUj-214	214	15 Dec 1939	9 Apr 1940	W/o Parola 2 Mar 1940	71.20
BUj-216	216	15 Dec 1939	20 Dec 1942	W/o Kauhava 9 Nov 1942	201.35

This table contains only those Bulldogs, which flew in the advanced training role, in the given period.

Bulldog BU-59 of LeSK at Kauhava on 19 February 1943, The yellow noses appeared late and only on a couple of Bulldogs, like BU-59 here. The upper surface camouflage is a solid Olive Green. (Finnish Air Force)

Bristol Bulldog IVA, BU-62, Lentolaivue 34, Parola, July 1940. Camouflage colours: upper surfaces Olive Green, under surfaces Light Grey, serial Black, tail White.

BU-62 of LLv 34, an advanced training squadron, at Parola on July 1940. It has a few months old factory camouflage of Olive Green top and Light Grey bottom. The White rudder indicated a trainer. (Erkki Komonen)

Bristol Bulldog IVA, BU-62, Lentolaivue 34, Parola, July 1940.

Bristol Bulldog IVA BU-68, Täydennyslentolaivue 35, Vesivehmaa, May 1942.
Camouflage colours: upper surfaces Olive Green, under surfaces Light Grey,
standard Eastern Front markings Yellow, serial Black, tail number White.

BU-68 of T-LeLv 35 parked at Vesivehmaa in May 1942.
The first flight flew four Bulldogs, BU-74 being number
2 and BU-67 number 3. The aircraft were handed over to
LeSK at the end of June 1942. (Olli Riekki)

Bristol Bulldog IVA, BU-71, Täydennyslentolaivue 35, Vesivehmaa, May 1942. Camouflage colours: upper surfaces Olive Green, under surfaces Light Grey, standard Eastern Front markings Yellow, serial Black, tail number White.

Bulldog BU-71 from the other side. Now with LeSK at Kauhava shortly before its crash on 13 October 1942. (Author's collection)

BU-71 of T-LeLv 35 at Vesivehmaa in May 1942. The mechanics are winding the inertia starter crank. All markings are fully by the book. (Olli Riekki)

Bristol Bulldog IVA BU-65, Lentosotakoulu, Kauhava, April 1943. Camouflage colours: upper surfaces Olive Green and Black, under surfaces Light Grey, standard Eastern Front markings Yellow, serial Black.

Bulldog BU-65 of LeSK at Kauhava in March 1943. It was the only Bulldog to receive the Warpaint, which was done at the factory on 10 July 1942. Being a trainer the undersides were painted Light Grey. (Author's collection)

Bristol Bulldog IVA, BU-65, Lentosotakoulu, Kauhava, April 1943.

113

Gloster Gauntlet II

Purchase

The Gauntlet was a British fighter design from 1929. It was built in two versions, a total of 245 examples. After the Winter War had started in late November 1939, the South African Federation supported Finnish aircraft acquisitions. They bought 29 Gauntlet IIs from Royal Air Force stocks and donated them to Finland. But only 24 ever arrived.

The first nine Gauntlets were shipped to Gothenburg, Sweden, where they were assembled and flown to Finland, the first two arriving on 10 March 1940 and the rest by 12 April 1940. The serials for these were GT-395–GT-403. The remaining fifteen were shipped to Finland, arriving by 17 May 1940. They were assembled by the air depot and given serials GT-404–GT-418.

Gauntlet II GT-402 after arrival in Finland at the air depot at Tampere on 12 April 1940. This was one of nine assembled in Sweden and, since the transfer flight required civil markings, the blue swastikas were overpainted. (Lassi Eskola collection)

GT-413 was shipped to Finland and assembled at the depot, pictured here at Tampere on 29 May 1940. On arrival all Gauntlet wore the late 1939 British fighter camouflage of Dark Green and Dark Earth upper surfaces. The lower surfaces were Black and White as demonstrated by the photo. (Finnish Air Force)

GT-416 after a belly landing repair at the factory at Tampere, shortly before delivery to LLv 34, which occurred on 3 June 1941. This plane was mostly in British colours, but the fuselage underside has received Light Grey paint. (VL)

GT-402 of T-LLv 25 on its nose at Ylivieska in early July 1941, still wearing the British colours. No report was found of this incident, but the damage looks light and repairable in the unit. (Author's collection)

Eight Gauntlets of T-LLv 25 moved from Vaasa to Vesivehmaa on 26 August 1941. They are here on a stop at Tampere, GT-414 being one them. It was repaired after a belly landing at the factory, which painted the plane on May 1941 olive Green on top and Light Grey underneath. (VL)

*Gauntlet GT-405 of
T-LLv 35 at Vesivehmaa
in March 1942. It had the
regulation trainer colours
of Olive Green upper and
Light Grey lower surfaces.
In all thirteen Gaunt-
lets received this type of
camouflage by May 1942.
(Author's collection)*

Employment

The first two were handed over in March 1940 to *T-LLv* 29, followed next spring by other advanced training squadrons *LLv* 30, *LLv* 34 and *T-LLv* 35.

At the beginning of the Continuation War on 25 June 1941 the remaining eleven Gauntlets were delivered to a temporary advanced training squadron, *LLv* 25. On 1 October 1941 they continued to *T-LLv* 17 and a month later to *T-LLv* 35. From late summer 1942 onwards the Gauntlets flew with *LeSK*, while one served with *T-LeLv* 17.

On 15 February 1945 the transfer began of the planes to storage at the air depot and over four days eleven Gauntlets were flown there, the last one being GT-396.

*GT-412 of T-LLv 35 hit the target in ground firing practices and bellied at Vesivehmaa on 1 July 1942. The solid Olive Green paintwork was just three months old.
(Olli Riekki)*

Colours

On arrival the Gauntlets wore the standard late 1939 British camouflage of Dark Green and Dark Earth upper surfaces with Black and White wing lower surfaces, the port ones being Black from the centre line.

In accident repairs or major overhauls at the State Aircraft Factory the Gauntlets received at first the Tuisku trainer type camouflage of Olive Green upper sides and in this case Light Grey undersides, since this was the factory practise for wood and fabric covered fighters. The first one to be painted this way was GT-395, in October 1940, followed by GT-418 and 416 later in 1940. The following year GT-415, 396, 398, 399, 414, 403 and 405 were successively repainted. In 1942 GT-412 and 410 followed suit.

GT-399 of T-LeLv 35 during a routine overhaul at the Mechanic School at Utti in September 1942. This plane was one of the early receivers of Warpaint, Olive Green and Black with DN-colour undersides, being applied on 15 July 1942. The Yellow noses appeared with the War-paint, though they were specified already on 1 September 1941 only for front-line fighters. (Finnish Air Force)

GT-402 of LeSK parked between the hangars at Kauhava in February 1943. It was the last to wear the solid Olive Green upper surfaces colour and kept it up to the crash on 7 July 1943. (Author's collection)

Outside this practise GT-416 received the Warpaint of Olive Green and Black in June 1941, and GT-417 likewise in May 1942, both retaining the Light Grey undersides.

Though the Gauntlet was never used operationally in Finland, it was classified as a fighter in May 1942 and subsequently received the standard Warpaint with Olive Green and Black tops and DN-colour undersides. During 1942 this was applied to GT-403, 411, 399, 408, 397, 418, 396, 414 and 400, in that order. In 1943 the Warpaint was applied to GT-406, 417, 401, 402 and 412. These colours were used until the planes were put into storage in February 1945.

Gauntlet GT-400 of T-LeLv 17 before take-off at Luonetjärvi on 1 April 1944, It was used by this advanced training squadron of bomber pilots for inspection flights. The Warpaint was applied on 2 October 1942. (SA-kuva)

GT-411 of LeSK on the platform at Kauhava in April 1943. This plane was the second to receive the full Warpaint, which consisted of Olive Green and Black upper surfaces and DN-colour undersides, painted on 7 July 1942. (Author's collection)

Serials

The Gauntlets wore the standard serials as specified on 20 March 1934, the height of the serial was 2½ times of the arm of the fuselage swastika. The nine planes (GT-395–GT-403) assembled in Sweden had the serials painted in a Swedish style font and the rest complied with the Finnish standard SFS Z.I.1. The Gauntlets assembled in Sweden also had the number of the serial in white on the black lower wing underside.

S/n	C/n	Delivered	Struck off charge	Remarks	Hours
GT-395		16 Mar 1940	10 Aug 1941	W/o Vaasa 19 Jul 1941	215.35
GT-396		14 Mar 1940	2 Jan 1950	Into storage 18 Feb 1945	339.35
GT-397		9 May 1940	2 Jan 1950	Into storage 16 Feb 1945	431.15
GT-398		9 May 1940	2 Jan 1950	Into storage 16 Feb 1945	281.05
GT-399		17 May 1940	2 Jan 1950	Into storage 16 Feb 1945	452.10
GT-400	K5271	13 Apr 1940	2 Jan 1950	Into storage 15 Feb 1945	133.35
GT-401		13 Apr 1940	2 Jan 1950	Into storage 16 Feb 1945	227.40
GT-402		13 Apr 1940	16 Sep 1943	W/o Munsala 7 Jul 1943	323.05
GT-403		13 Apr 1940	10 Oct 1942	W/o Alahärmä 12 Aug 1942	402.35
GT-404	7	14 Jun 1940	23 Apr 1941	W/o Ulvila 4 Jan 1941	101.50
GT-405	8	14 Jun 1940	2 Apr 1943	W/o Lapua 15 Dec 1942	341.10
GT-406	9	5 Jun 1940	2 Jan 1950	Into storage 15 Feb 1945	269.25
GT-407	11	12 Jun 1940	23 Apr 1941	W/o Ulvila 4 Jan 1941	166.30
GT-408	14	5 Sep 1940	6 Nove 1944	Into storage 26 Aug 1944	455.10
GT-409	15	5 Jun 1940	29 Aug 1940	W/o Ulvila 18 Jul 1940	6.20
GT-410	17	24 Apr 1942	9 Jan 1943	W/o Kruunukylä 30 Sep 1942	79
GT-411	18	5 Jun 1940	6 Nov 1944	Into storage 22 Jul 1944	388.05
GT-412	19	30 May 1940	2 Jan 1950	Into storage 15 Feb 1945	216.50
GT-413	21	30 May 1940	8 Oct 1940	W/o Hattula 18 Sep 1940	199.40
GT-414	22	10 Jun 1940	2 Jan 1950	Into storage 17 Feb 1945	521.35
GT-415	23	29 May 1940	6 Nov 1941	W/o Kokemäki 28 Mar 1941	133.55
GT-416	16	12 Jun 1940	2 Jan 1950	Into storage 15 Feb 1945	353.30
GT-417	20	5 Jun 1940	2 Jan 1950	Into storage 15 Feb 1945	405.05
GT-418	24	5 Jun 1940	6 Nov 1944	W/o Mustasaari 4 Jul 1944	229.50

GT-398 of LeSK at the ramp at Kauhava in May 1944. In the rare case of enemy air attacks on Kauhava, the school's Gauntlets were in readiness to intercept, as they were the only armed aircraft at the time. (Author's collection)

Gloster Gauntlet II, GT-396, Täydennyslentolaivue 35, Parola, June 1940. Camouflage colours: upper surfaces Dark Green and Dark Earth, under surfaces Black and White, serial Black, rudder White.

After arrival in Finland on 10 March 1940, GT-396 went straight to T-LLv 29, which was based at Tyrväntö. This machine was assembled in Sweden, where the serial was painted with Swedish style font. Also the lower wing Black undersides had large White 396 digits. (Author's collection)

The engine of GT-396 of T-LLv 35 quit and the plane was damaged at Parola on 12 June 1940. The white rudder identified trainers. The plane was repaired and flew again eight months later. (Eino Ritaranta collection)

Gloster Gauntlet II, GT-396, Täydennyslentolaivue 35, Parola, June 1940.

Gloster Gauntlet II, GT-408, Täydennyslentolaivue 25, Vaasa, August 1941. Camouflage colours: upper surfaces Olive Green, under surfaces Light Grey, standard Eastern Front markings Yellow, serial Black, tail number White.

GT-408 of T-LLv 408 on a stop at Tampere on 26 August 1941, on the way to the unit's new base at Vesivehmaa. Only a couple of this advanced squadron's Gauntlets had a tactical fighter type tail number. (VL)

Gloster Gauntlet II, GT-416, Lentosotakoulu, Kauhava, March 1944. Camouflage colours: upper surfaces Olive Green and Black, lower DN-colour, standard Eastern Front markings Yellow, serial Black, tail number White.

Gloster Gauntlet II, GT-416 in March 1944. At right it is seen after a light mid-air collision with GT-418 on 13 March 1944, both planes making a normal landing at

GT-416 of LeSK at Kauhava in March 1944. At right it is seen after a light mid-air collision with GT-418 on 13 March 1944, both planes making a normal landing at Kauhava. (Finnish Air Force Museum)

Gloster Gauntlet II, GT-414, Lentosotakoulu, Kauhava, March 1943. Camouflage colours: upper surfaces Olive Green and Black, lower DN-colour, standard Eastern Front markings Yellow, serial Black.

Gauntlet GT-414 of LeSK photographed in early 1943, at left at Kauhava on 19 February 1943 and at right at Lappajärvi camp on 10 March 1943. The Warpaint was applied on 19 August 1942. (Finnish Air Force)

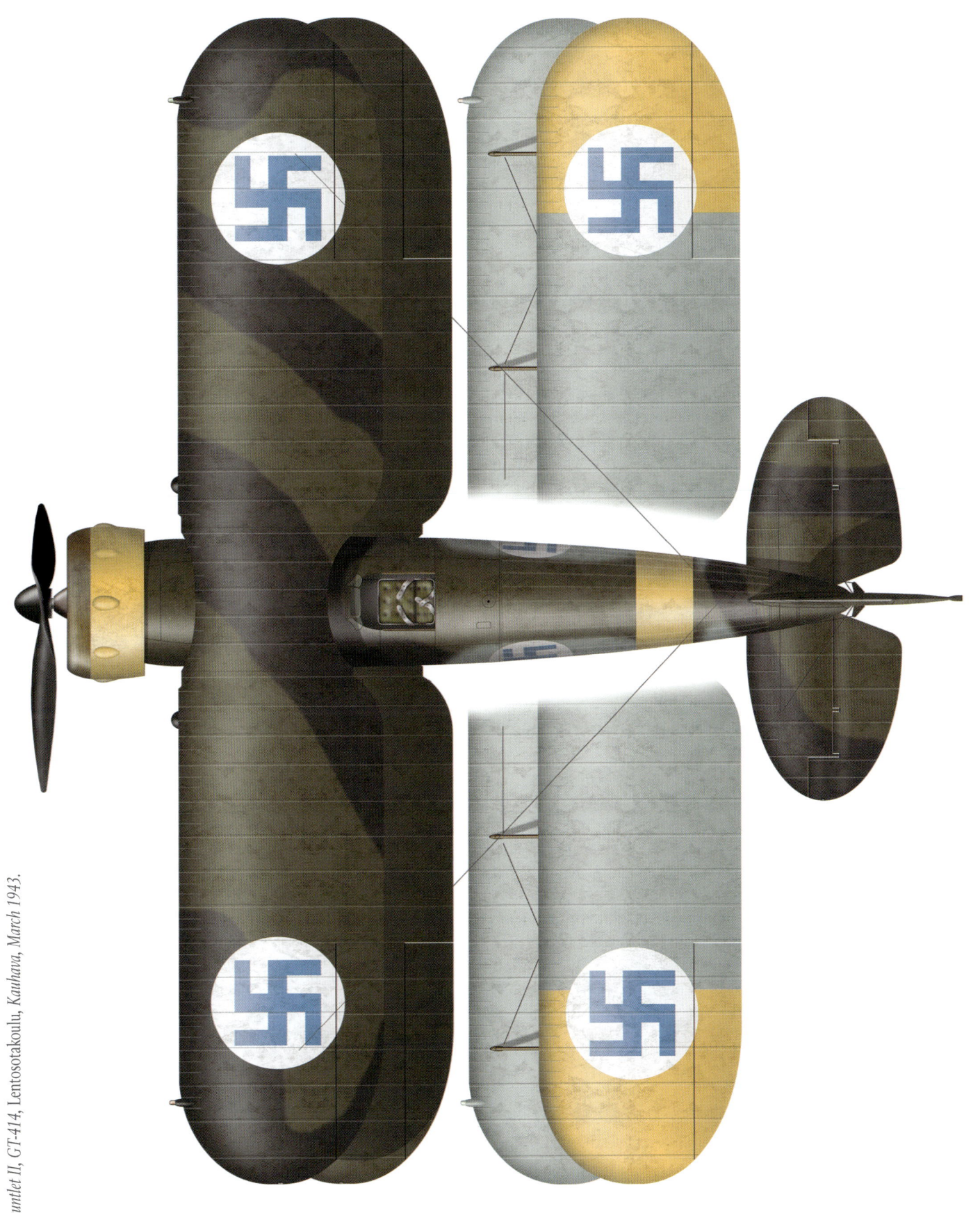

Gloster Gauntlet II, GT-414, Lentosotakoulu, Kauhava, March 1943.

Polikarpov I-15bis

Purchase

During the Winter War in 1939–40 a number of I-15bis fighters made forced landings and were captured. Five were later refurbished into flying condition. The first two were coded VH-10 and VH-11 according to the marking instructions issued on 12 February 1940. The next three were serialled VH-3 to VH-5. At the end of 1940 VH-10 and 11 were respectively re-numbered as VH-1 and 2, to avoid overlapping of serials with the I-153 type of aircraft. On 4 June 1942 the serial problem was finally solved and I-15bis aircraft were coded IH-1–IH-5.

Employment

The second of five captured I-15bis fighters refurbished into flying condition was VH-11. It is photographed here at the State Aircraft Factory at Tampere in February 1940. The VH letters in the serial come from Venäläinen Hävittäjä (Russian Fighter). The camouflage is the standard olive Green upper sides and Light Grey lower sides introduced in January 1934. (VL)

During the Winter War, out of five I-15bis aircraft two made it in time to squadron service with *T-LLv* 29, which was an advanced training unit.

Between the wars all others we refurbished at the factory and they were placed in *LLv* 34 as fighter trainers. When *T-LLv* 29 was disbanded at the end of March 1940, its equipment was transferred to *LLv* 34.

At the beginning of the Continuation War *LLv* 34 was de-activated and the I-15bis planes were transferred to *T-LLv* 35 as trainers. In autumn 1942 two aircraft were put into storage. In summer 1943 three I-15bis aircraft served in the re-activated *LeLv* 34 on target and target-tow duties. During the next autumn and winter *LeLv* 30 still had two planes.

In 1944 two aircraft served with *T-LeLv* 35. After the war the aircraft were stored at the air depot. The last flight of the type in the Finnish Air Force was made on 12 March 1945, when IH-4 and IH-5 were flown to the depot.

VH-1 of T-LLv 35 in Vesivehmaa hangar in May 1942. Its earlier serial was VH-10 and the later one was IH-1. The camouflage is the standard solid Olive Green tops and Light Grey bottoms. (Olli Riekki)

Colours

All five repaired aircraft were given the contemporary regulation camouflage of Olive Green upper surfaces and Light Grey undersides. Following accident repairs, three I-15bis aircraft were painted in the Warpaint of Olive Green and Black with DN-colour undersides. The first one to receive this was IH-3 in October 1942. IH-2 followed a month later and IH-4 was the last in February 1943.

IH-4 of T-LeLv 35 at Kauhava in July 1944. It wears the full Warpaint of Olive Green and Black upper surfaces with DN-colour lower surfaces and standard Yellow Eastern Front markings, including the lower half of the nose. The upper half Yellow was deleted from 13 June 1944 onwards. (Aarno Kaila)

S/n	Delivered	Struck off charge	Remarks	Hours
VH-10, VH-1, IH-1	23 Feb 1940	2 Jan 1950	Into storage 11 Sep 1942	109.05
VH-11, VH-1, IH-2	26 Feb 1940	2 Jan 1950	Into storage 20 Feb 1945	114.50
VH-3, IH-3	19 Sep 1940	2 Jan 1950	Into storage 20 Feb 1945	115.25
VH-4, IH-4	11 Oct 1940	2 Jan 1950	Into storage 12 Mar 1945	144.25
VH-5, IH-5	3 Feb 1941	2 Jan 1950	Into storage 12 Mar 1945	106.55

Polikarpov I-15bis, VH-2, Täydennyslentolaivue 35, Vesivehmaa, May 1942. Camouflage colours: upper surfaces Olive Green, under surfaces Light Grey, standard Eastern Front markings Yellow, serial Black.

VH-2 shortly after arrival at Vesivehmaa to T-LLv 35, on 17 October 1941. The tail is lifted for gun harmonization purposes. Yellow paint was applied under both wing tips. (Author's collection)

VH-2 of T-LLv 35 at Vesivehmaa in February 1942, fitted with original Russian skis. There is no Yellow on the nose. It appeared with the Warpaint on 11 November 1942. (Olli Riekki)

Polikarpov I-15bis, IH-2, Lentolaivue 34, Utti, July 1943. Camouflage colours: upper surfaces Olive Green and Black, under surfaces DN-colour, standard Eastern Front markings Yellow, serial Black.

IH-2 was the target and target-tow aircraft of LeLv 34, arriving at Utti on 17 July 1943, but becoming a write-off only two weeks later. It has the regulation Warpaint. (Paavo Saari)

VL *Pyry* I

Purchase

As an extension to a number of biplanes designed and built by the State Aircraft Factory, the air force ordered the design of a monoplane advanced trainer on 23 October 1937 and later a prototype on 10 December 1938. Dipl. Ing. Arvo Ylinen was appointed as the chief designer and the prototype was named *Pyry*. It was built by early 1939. The first flight was performed on 29 March 1939, the prototype bearing serial PY-1.

The *Pyry* met the requirements, apart from slight stall and balance problems, and the factory received on 3 May 1939 an order to manufacture 40 aircraft. These were given serials PY-2–PY-41 and they were built between December 1940 and June 1941. Thus the Finnish Air Force had a total of forty-one *Pyrys* in its inventory.

Pyry prototype with serial PY-1 at the factory a Tampere in July 1939. The horn balanced rudder is new. The camouflage is the regulation Olive Green and Light Grey. (VL)

Pyry PY-3 of LLv 34 at Hyvinkää in May 1941. All production Pyrys received the 30 September 1940 ordered camouflage called Warpaint, Olive Green and Black upper surfaces and Light Grey lower surfaces. The latter was for aircraft with wooden or fabric skinning. The serial was Green on the Black areas. (Author's collection)

The type showed slightly unsatisfactory stall characteristics, caused by the elliptical wing. Efforts to correct this were made by installing, on 30 March 1941, a tapered wing to *Pyry* PY- 24 and later to PY-1, 32 and 37, but without any significant improvements. All but PY-24 reverted back to the elliptical wing. The stall problem found its solution in slotted wings, which were installed from 4 December 1941 onwards to *Pyrys*.

The balance problem flying with a two-man crew was solved on 10 April 1943 by extending the engine mount by 10 cms. All these modifications were done at the next major overhaul or repair.

Pyry PY-16 under evaluation with LLv 32 at Utti in July 1941. The colours and markings are by the book. The wheel spats acted as a canvas for creative artwork. (Kyösti Karhila)

PY-16 now as an advanced trainer with T-LLv 25 at Mustasaari near Vaasa in August 1941. The yellow fuselage band has a seldom-seen cutout at the serial number. (Olli Riekki)

Pyry PY-29 of T-LLv 25 at Mustasaari in August 1941, wearing the factory applied Warpaint of Olive Green and Black. (Olli Riekki)

PY-22 of LeSK at Kauhava in October 1941. On this plane the Yellow fuselage band also has a cut-out at the serial. (Author's collection)

PY-33 of LeSK photographed on the first snow at Kauhava on 23 October 1941. The wheel spats have been removed to prevent snow clogging the wheels. (Finnish Air Force)

Employment

After the test flight the prototype PY-1 was handed over to T-LLv 29 on 20 December 1939 as an advanced trainer. After this unit was disbanded PY-1 was transferred on 1 June 1940 to *LLv* 34, another advanced training squadron.

From January 1941 onwards the series machines entered service with *LeSK* and *LLv* 34. Additionally all operational squadrons had two or three *Pyrys* for evaluation purposes.

At the beginning of the Continuation War in late June 1941 the squadron machines were delivered to a new and temporary advanced training squadron, *T-LLv* 25. *LeSK* continued to fly the *Pyry* and likewise *LLv* 34.

On 1 October 1941 *T-LLv* 25 and *LLv* 34 were merged into *T-LLv* 35. At this point some of the *Pyrys* were handed over to *T-LLv* 17, which was in charge of reconnaissance and bomber aircrew advanced training. A month later *T-LLv* 17 was disbanded and *LLv* 46, and a month later *LLv* 48, took over its responsibilities.

From early August 1942 onwards the *Pyrys* served in only two units, *LeSK* and *T-LeLv* 35, apart from solitary liaison aircraft in some front-line squadrons.

PY-22 of LeSK on its nose at Kauhava on 14 January 1942, showing clearly the standard camouflage pattern. Landing with the flaps up caused the plane to overshoot the runway and finish up on its nose. (Author's collection)

In early August 1944 *LeSK* became the sole user of the *Pyry*. The Continuation War ended on 4 September 1944 in an armistice and two weeks later came the Moscow Truce. At this point the Allied Supervision Commission grounded all aircraft except those which were sent to drive off the Germans in Lapland.

Peace time flying was resumed in early August 1945 and the main *Pyry* user was *LeSK*, which was named *IlmaSK* (Air Fighting School) on 1 December 1952. The *Pyry* continued to serve until September 1961 and the last flight was performed by PY-27 on 10 September 1962.

PY-6 of T-LeLv 35 on its nose at Vesivehmaa on 27 August 1942. The rudder wears a Blue and White number 1. (Author's collection)

PY-4 liaison of LeLv 28 flipped over at Viitana on 25 July 1942, and has now been righted. An ace of spades decorates the wheel strut cover. (Author's collection)

PY-29 of T-LLv 35 at Vesivehmaa in March 1942. Hard use begins to show on the paintwork of the plane. (Olli Riekki)

Colours

The *Pyry* prototype PY-1 and the first series machine PY-2 were finished in the standard warplane camouflage established in January 1934, olive green upper surfaces. Instead of aluminium dope lower surfaces, the factory had introduced a Light Grey. The Warpaint became effective on 30 September 1940 and PY-1 was painted in this Olive Green and Black scheme in March 1941 and PY-2 in June 1941.

All subsequent *Pyrys* received Warpaint as the factory finish with light grey undersides. When the Light Blue-Grey (RLM 65) lower surface DN-colour was introduced on 7 May 1942, it concerned only warplanes. But *Pyry* advanced trainers made an exception and received DN-colour undersides in the next major overhaul or repair. In 1942 only PY-8 got this in October. In 1943 eleven *Pyrys* received the DN-colour and next year another dozen.

The warpaint existed on *Pyrys* until its removal on 27 September 1947. In a repair or major overhaul after that, *Pyrys* were painted in the standard trainer colours of Olive Green fuselage and orange flying surfaces.

PY-39 of T-LeLv 35 at Vesivehmaa in August 1942. The white rudder indicates a trainer. Fuel spill has removed most of the paint on the upper decking. (Author's collection)

PY-11 of LeSK at Lappajärvi shortly before it was damaged in a landing on 20 March 1943. This view shows clearly the wing slots, which eliminated the small stall problem. (Author's collection)

PY-13 of T-LeLv 35 paying a visit to Helsinki Malmi in autumn 1943. The Blue and White tail number indicated the first flight. (Esko Rinne)

Serials

The *Pyrys* wore the standard serials as specified on 20 March 1934, the height of the serial was 2½ times of the arm of the fuselage swastika. The style complied with the Finnish standard SFS Z.I.1.

PY-31 of LeSK at Kauhava, where it arrived on 6 March 1944. This machine got the DN-colour bottoms on 16 June 1943. (Lars Bergman)

PY-31 still with T-LeLv 35 at Vesivehmaa in February 1944. The white crosses on the wings have been painted for aiming purposes. (Pentti Manninen collection)

S/n	C/n	Delivered	Struck off charge	Remarks	Hours
PY-1		20 Dec 1939	18 Sep 1962	Last flight 1 Sep 1962	2222.25
PY-2	I/1	9 Jan 1941	27 Sep 1947	W/o Vanajavesi 26 Aug 1947	848.35
PY-3	I/2	29 Jan 1941	29 Apr 1942	W/o Hyvinkää 15 Jun 1941	74.15
PY-4	I/3	8 Feb 1941	27 Apr 1960	W/o Kauhava 21 Mar 1959	1156.45
PY-5	I/4	8 Feb 1941	18 Dec 1958	W/o Kauhava 10 Jul 1957	2308.10
PY-6	I/14	18 Mar 1941	12 Dec 1943	W/o Asikkala 9 Sep 1943	447.40
PY-7	I/5	19 Feb 1941	16 Mar 1957	Last flight 5 Dec 1956	2589.45
PY-8	I/6	25 Feb 1941	16 Oct 1961	W/o Kauhava 3 Jul 1961	2543.20
PY-9	I/7	25 Feb 1941	22 Feb 1943	W/o Kauhava 15 Jan 1943	536,25
PY-10	I/8	26 Feb 1941	1 Jul 1949	W/o Vaasa 9 Jun 1949	1269.40
PY-11	I/9	27 Feb 1941	16 Mar 1957	W/o Malmi 23 Sep 1956	1979.40
PY-12	I/10	27 Feb 1941	4 Apr 1951	W/o Parkano 2 Mar 1951	1068.50
PY-13	I/11	11 Mar 1941	16 Mar 1957	Last flight 26 Aug 1955	2204.45
PY-14	I/12	18 Mar 1941	10 Jan 1944	W/o Vesivehmaa 17 Dec 1943	355.40
PY-15	I/13	18 Mar 1941	29 Apr 1942	W/o Vesijärvi 10 Mar 1942	288.30
PY-16	I/15	27 Mar 1941	2 Jun 1962	W/o Orivesi 22 Feb 1962	2482.05
PY-17	I/16	28 Mar 1941	22 Jun 1950	W/o Lappajärvi 31 Mar 1950	1274.05
PY-18	I/17	28 Mar 1941	26 Apr 1944	W/o Asikkala 31 Mar 1943	718.15
PY-19	I/18	28 Mar 1941	16 Mar 1957	W/o Kauhava 17 Aug 1956	3016.30
PY-20	I/19	3 Apr 1941	16 Mar 1957	Into storage 7 Nov 1956	2127.55
PY-21	I/20	2 Apr 1941	24 Nov 1944	W/o Kauhava 21 Jul 1944	698.55
PY-22	I/21	2 Apr 1941	16 Aug 1943	W/o Vesivehmaa 31 May 1943	578.45
PY-23	I/24	4 Apr 1941	16 Mar 1957	W/o Luonetjärvi 29 May 1956	2278.55
PY-24	I/22	3 Apr 1941	27 Apr 1960	W/o Utti 16 Apr 1960	2580.20
PY-25	I/23	7 Apr 1941	22 May 1943	W/o Utti 27 Mar 1943	410.15
PY-26	I/27	19 Apr 1941	2 Jun 1962	Last flight 28 Jul 1960	2243.55
PY-27	I/25	23 Apr 1941	18 Sep 1962	Last flight 10 Sep 1962	2271.05
PY-28	I/26	15 Apr 1941	18 Dec 1958	W/o Kauhava 28 Mar 1958	2365.05
PY-29	I/28	15 Apr 1941	21 Sep 1948	W/o Kuorevesi 9 Apr 1948	709.45
PY-30	I/29	24 Apr 1941	16 Mar 1957	W/o Malmi 20 Apr 1956	1867
PY-31	I/30	29 Apr 1941	16 Mar 1957	W/o Pori 20 Apr 1956	2459.25
PY-32	I/31	29 Apr 1941	13 Apr 1944	W/o Lempäälä 19 Feb 1944	451.35
PY-33	I/32	29 Apr 1941	26 Apr 1944	W/o Asikkala 31 Mar 1944	680
PY-34	I/33	29 Apr 1941	29 Apr 1942	W/o Vesijärvi 10 Mar 1942	194.10
PY-35	I/34	30 Apr 1941	2 Aug 1962	Last flight 14 Jul 1962	1999
PY-36	I/35	3 Jun 1941	16 Mar 1957	Last flight 28 Jan 1956	1755.10
PY-37	I/36	3 Jun 1941	22 May 1943	W/o Tampere 7 Mar 1943	290
PY-38	I/38	4 Aug 1941	16 Aug 1943	W/o Vesivehmaa 18 Jun 1943	394
PY-39	I/40	1 Sep 1941	10 Oct 1942	W/o Juupajoki 12 Aug 1942	211.30
PY-40	I/37	4 Aug 1941	24 Nov 1944	W/o Kauhava 21 Jul 1944	716.25
PY-41	I/39	4 Aug 1941	27 Apr 1960	Last flight 4 Mar 1958	2020.10

The last Pyry built was this PY-41, here with LeSK at Kauhava in March 1944. The brand new Warpaint with DN-colour unbdersides was applied on 26 February 1944. (Aarno Kaila)

VL Pyry prototype, PY-1, Valtion Lentokonetehdas, Tampere, July 1939. Camou-flage colours: upper surfaces Olive Green, under surfaces Light Grey, serial Black.

Pyry prototype PY-1 at the factory at Tampere in July 1939, after small modifications, including the horn balanced rudder. (VL)

VL Pyry I, PY-5, Lentosotakoulu, Kauhava, June 1943. Camouflage colours: upper surfaces Olive Green and Black, under surfaces DN-colour, standard Eastern Front markings Yellow, serial Black.

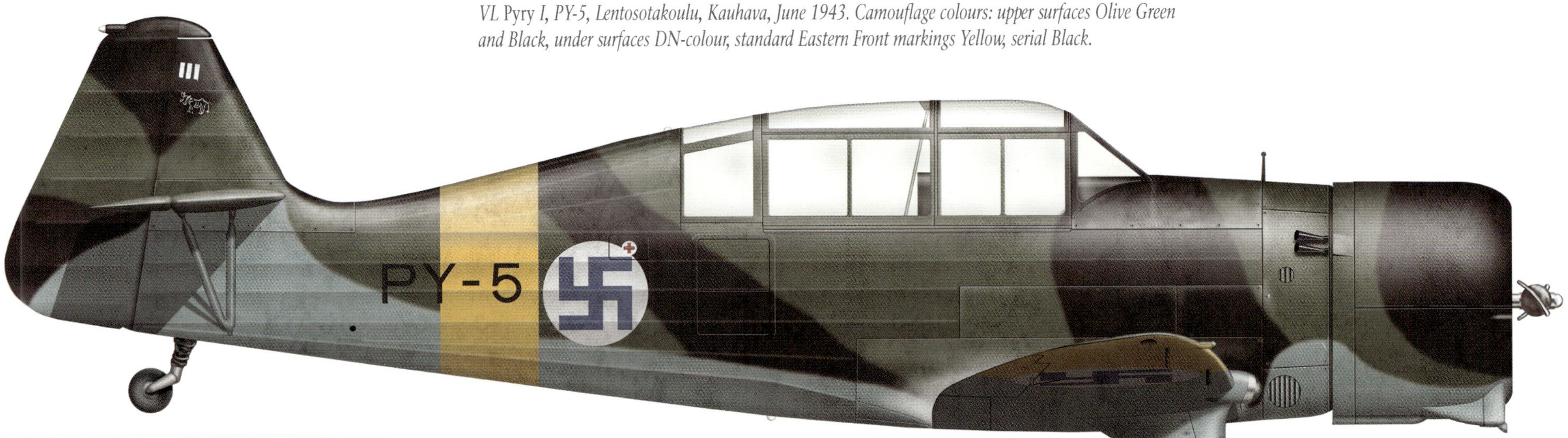

VL Pyry I, PY-5 of LeSK at Kauhava in June 1943. Showing fighter pilot spirit, the fin has three kill bars, probably from scaring cows of a nearby farm. This was the third Pyry to receive the DN-colour bottom on 27 March 1943. (Finnish Air Force Museum)

VL Pyry I, PY-18, Lentosotakoulu, Kauhava, July 1941. Camouflage colours: upper surfaces Olive Green, Black and Light Grey, under surfaces Light Grey, standard Eastern Front markings Yellow, serial Black (port) and Olive Green (starboard).

PY-18 of LeSK at Kauhava on July 1941 (right) and a few months later. Quite exceptionally the light grey "clouds" extend far on the upper surfaces of the wings. Also the Yellow fuselage band is in a more forward position. (Author's collection)

140

VL Pyry I, PY-18, Lentosotakoulu, Kauhava, July 1941.

VL Pyry I, PY-24, Täydennyslentolaivue 35, Vesivehmaa, June 1943. Camouflage colours: upper surfaces Olive Green and Black, under surfaces Light Grey, standard Eastern Front markings Yellow, serial Olive Green (starboard), tail number White.

Five Pyrys of T-LeLv 35 in a line at Vesivehmaa on 5 June 1943. The planes are, from the left, PY-24, 26 and 20. PY-24 was the first fitted with a tapered wing in hopes for better stall characteristics, but it did not work. (SA-kuva)

VL Pyry I, PY-24, Täydennyslentolaivue 35, Vesivehmaa, June 1943.

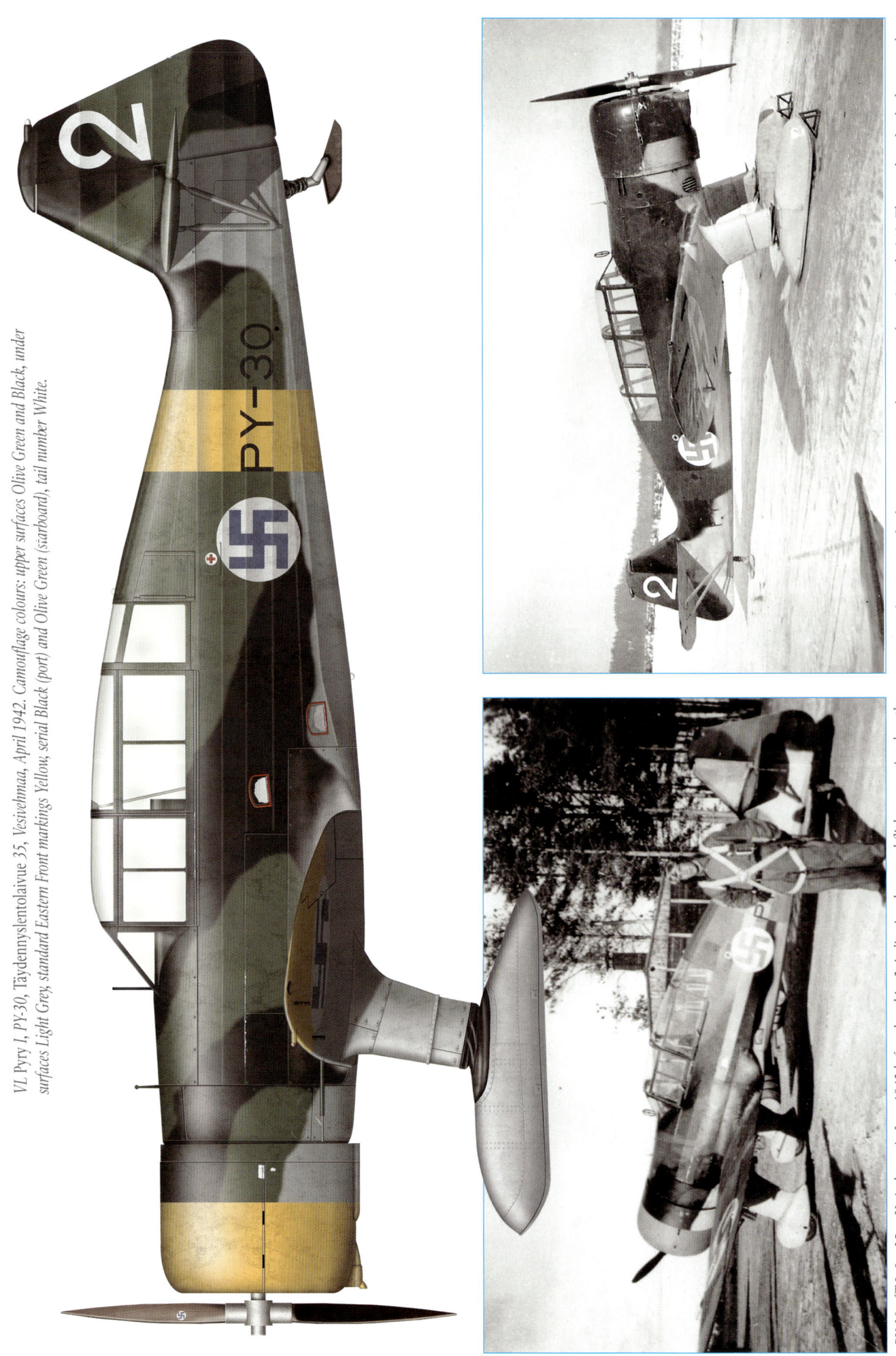

VL Pyry I, PY-30, Täydennyslentolaivue 35, Vesivehmaa, April 1942. Camouflage colours: upper surfaces Olive Green and Black, under surfaces Light Grey, standard Eastern Front markings Yellow, serial Black (port) and Olive Green (starboard), tail number White.

PY-30 as a liaison aircraft of LeLv 34 at Utti, where it arrived on 12 April 1943. The stay lasted only three months. (Esko Laiho)

PY-30 of T-LeLv 35 at Vesivehmaa in June 1942 before the white 2, indicating the second flight, was painted on the rudder. This plane is the only known Pyry to have a Yellow nose, intended for fighters only. (Olli Riekki)

VL Pyry I, PY-40, Täydennyslentolaivue 17, Luonetjärvi, March 1943. Camouflage colours: upper surfaces Olive Green and Black, under surfaces Light Grey; standard Eastern Front markings Yellow, serial Black (port), tail number Blue and White.

PY-40 of T-LeLv 17 in the Luonetjärvi hangar in March 1943. It came from T-LeLv 35 still bearing its fighter-type tail number. (Author's collection)

145

Fokker D.XXI

Purchase

Finland was the first customer to buy the Dutch Fokker D.XXI, when on 18 November 1936 an order was placed for seven aircraft and a licence to manufacture 14 more. The Dutch-built aircraft were shipped to Finland in November 1937. These I series Fokkers were serialled FR-76–FR-82.

The fourteen licence-built aircraft were ordered from the State Aircraft Factory on 7 May 1937. This II series were completed between November 1938 and March 1939 and serialled FR-82–FR-96.

On 15 June 1937 an unlimited licence was acquired and twenty-one planes were ordered from the factory. The III series machines were assembled between March and July 1939 and coded FR-97–FR-117.

On 9 May 1940 a fifty plane IV series was ordered. Since the Mercury power plants went to the Bristol Blenheim licence production, the aircraft were fitted with a 825 hp R-1535 Twin Wasp Jr engines. The Fokkers were built between October 1940 and June 1941 and serialled FR-118–FR-167.

During summer 1944 a five plane V series was constructed from spares and serialled FR-171–FR-175. These were also powered with the Twin Wasp.

The Finnish Air Force had in its inventory a total of 97 Fokker D.XXIs, 55 of these being Twin Wasp-powered.

Fokker D.XXI coded FR-139 of T-LeLv 35 at Vesiveh-maa before an accident on 28 May 1942. The Warpaint of Olive Green and Black with Light Grey undersides was applied at the factory. The white rudder indicated a trainer. (Olli Riekki)

FR-143 of T-LeLv 35 on gunnery practice at Joensuu in July 1942. The White bar on the rudder indicated a section of UK 13 (officer course 13). The camouflage is the regulation Warpaint. (Finnish Air Force)

Employment

From December 1937 the first phase users were fighter squadrons *LLv* 24 and *LLv* 26, both fighting in the Winter War. Thereafter fighter squadrons *LLv* 30 and *LLv* 32 flew the type early in the Continuation War. In the second phase the D.XXI was operated by reconnaissance squadrons *LLv* 12 and *LLv* 14, practically throughout the Continuation War.

The D.XXI entered the training role on 2 April 1941, when two aircraft were delivered to *LLv* 34, which was the training element of *LeR* 3. On 25 June 1941 *LLv* 34 was subordinated to *T-LLv* 35 and the latter became, on 1 October 1941, the sole advanced training squadron for fighter and reconnaissance pilots. *T-LLv* 35 received five more D.XXIs by the end of 1941.

For the rest of the Continuation War *T-LeLv* 35 had an average of ten D.XXIs serviceable. Due to attrition the flow of replacement planes kept going. In all *T-LeLv* 35 carried out 30 FR-courses.

After the Moscow truce on 14 September 1944, all trainers were grounded and on 27 November 1944 *T-LeLv* 35 was disbanded and annexed to *LeSK*. The D.XXIs at *LeSK* resumed flying in early August 1945 and flew until 13 September 1948, when FR-139 and -174 performed the last flights of the type.

FR-152 of T-LeLv 35 also at the gunnery practices at Joensuu in July 1942. The factory applied camouflage is almost identical to FR-143. This plane has two White bars on the rudder. (Finnish Air Force)

FR-152 of T-LeLv 35, fresh from repair at the factory, at Vesivehmaa in June 1943. The colours and markings are still quite similar, except now the lower surfaces are finished in DN-colour. (Author's collection)

Colours

Except the prototype FR-118, the factory finish, for the Wasp-powered D.XXIs was the Warpaint of Olive Green and Black with Light Grey undersides. Being classified as a warplane the D.XXIs began receiving the Light Blue-Grey DN-colour lower surfaces from 7 May 1942 onwards, in the next repair or major overhaul.

The first plane to be painted in these colours was FR-157 on 1 June 1942. The next eight during 1942 being chronologically FR-162, 132, 133, 154, 134, 139, 160 and 125. In 1943 a further fourteen aircraft received the DN-väri underside and in 1944 yet another four. These colours remained on the planes to the end of their flying career.

FR-171 of T-LeLv 35 arrived at Kauhava on 7 August 1944. It was in the last batch of five Fokkers which the factory built, wearing typical 1944 camouflage and markings. (Aarno Kaila)

S/n	C/n	Delivered	Struck off charge	Remarks	Hours
FR-83	II/1	8 Aug 1944	1 Oct 1952	Into storage 23 Feb 1945	
FR-92	II/10	8 Aug 1944	1 Oct 1952	Into storage 23 Feb 1945	
FR-98	III/2	8 Aug 1944	1 Oct 1952	Into storage 26 Feb 1945	
FR-100	III/4	8 Aug 1944	1 Oct 1952	Into storage 23 Feb 1945	
FR-108	III/9	8 Aug 1944	1 Oct 1952	Into storage 26 Feb 1945	
FR-120	IV/4	23 Aug 1941	6 Nov 1941	W/o Vesivehmaa 9 Oct 1941	
FR-121	IV/5	23 Jul 1943	3 Mar 1944	W/o Vesivehmaa 18 Jan 1944	
FR-122	IV/3	4 Dec 1941	2 Jan 1950	W/o Kauhava 26 Sep 1946	
FR-125	IV/8	22 Jan 1943	22 Jun 1950	Into storage 26 Feb 1945	
FR-129	IV/12	24 Feb 1944	1 Oct 1952	Into storage 24 Feb 1945	
FR-130	IV/13	23 Nov 1942	1 Oct 1952	Into storage 23 Feb 1945	
FR-131	IV/14	7 Aug 1943	1 Oct 1952	Into storage 2 Nov 1945	
FR-132	IV/15	2 Apr 1941	1 Oct 1952	Into storage 26 Mar 1945	
FR-133	IV/16	2 Apr 1941	25 Sep 1944	W/o Suurjärvi 27 Aug 1944	
FR-134	IV/17	19 Sep 1942	30 Aug 1944	W/o Härmä 12 Jul 1944	
FR-135	IV/18	2 Aug 1943	1 Oct 1952	Into storage 26 Jan 1946	
FR-136	IV/19	30 Sep 1942	1 Oct 1952	Into storage 13 Sep 1948	
FR-139	IV/22	1 Nov 1941	1 Oct 1952	Into storage 13 Sep 1948	
FR-141	IV/24	8 Feb 1943	16 Aug 1943	W/o Hollola 20 May 1943	
FR-142	IV/25	25 Jul 1944	27 Feb 1953	Into storage 13 Sep 1948	
FR-143	IV/26	3 Dec 1941	1 Oct 1952	Into storage 16 Feb 1945	
FR-144	IV/27	21 Aug 1943	1 Oct 1952	Into storage 19 Sep 1945	
FR-145	IV/28	27 May 1944	1 Oct 1952	Into storage 13 Sep 1948	
FR-146	IV/29	1 Jun 1944	1 Oct 1952	Into storage 17 Feb 1945	
FR-152	IV/35	4 Dec 1941	1 Oct 1952	Into storage 17 Feb 1945	
FR-156	IV/39	17 Feb 1944	1 Oct 1952	Into storage 23 Feb 1945	
FR-162	IV/45	26 May 1944	30 Aug 1944	W/o Kauhava 27 Jun 1944	
FR-167	IV/50	11 Jun 1942	1 Oct 1952	Into storage 13 Sep 1948	
FR-171	V/1	7 Aug 1944	27 Feb 1953	W/o Kauhava 30 Jul 1948	
FR-172	V/3	20 Aug 1944	27 Fen 1953	Into storage 13 Sep 1948	
FR-173	V/2	14 Aug 1944	27 Feb 1953	Into storage 13 Sep 1948	
FR-174	V/4	14 Aug 1944	1 Oct 1952	Into storage 13 Sep 1948	

This table contains only those Fokker D.XXIs which flew in the advanced training role in the given period.

FR-156 of T-LeLv 35 ready for a training sortie at Kauhava in August 1944. The white tail number referred to planes in the second flight, which at that time had more than ten Fokkers. (Finnish Air Force Museum)

FR-141 of T-LeLv 35 at Vesivehmaa in May 1943. In training use the planes were fitted with a State Aircraft Factory designed and built wooden propellers, saving the constant-speed metal propellers for combat use. (Author's collection)

Half a dozen Fokker D.XXIs of T-LeLv 35 inside the hangar at Vesivehmaa on 4 June 1943. The nearest machine, FR-134, has a different trainer sign, a White fin. T-LeLv 35 had an average of ten D.XXIs in working order. (SA-kuva)

FR-167 of T-LeLv 35 at Vesivehmaa in May 1943. This was one of two D.XXIs with a Finnish designed retractable landing gear. It collapsed in a landing on 22 May 1944 and was converted back to a fixed gear. (Author's collection)

On 8 August 1944 T-LeLv 35 received at Kauhava the five remaining Mercury-powered D.XXIs. Here is FR-83 in standard Warpaint, but with DN-colour underside and subdued national insignias. (Aarno Kaila)

Fokker D.XXI, FR-143, Täydennyslentolaivue 35, Joensuu, July 1942. Camouflage colours: upper surfaces Olive Green and Black, under surfaces Light Grey, standard Eastern Front markings Yellow, serial Black, tail stripe White.

FR-143 of T-LeLv 35 in gunnery practice held at Joensuu in July 1942. Later these practices were carried out over Lake Vesijärvi next to Vesivehmaa. (Finnish Air Force)

Fokker D.XXI, FR-133, Täydennyslentolaivue 35, Vesivehmaa, August 1942.
Camouflage colours: upper surfaces Olive Green and Black, under surfaces
DN-colour, standard Eastern Front markings Yellow, serial Black, rudder White.

FR-133 of T-LeLv 35 parked at the fuel pump at Utti in
August 1942. It has just come out of a regular service performed by the Mechanic School based there. (Paavo Saari)

Fokker D.XXI, FR-132, Täydennyslentolaivue 35, Vesivehmaa, October 1942. Camouflage colours: upper surfaces Olive Green and Black, under surfaces DN-colour, standard Eastern Front markings Yellow, serial Black, fin White.

FR-132 of T-LeLv 35 after a hard landing at Vesivehmaa on 26 October 1942. The White fin was one of several signs of an advanced trainer. (Author's collection)

Fokker D.XXI, FR-139, Täydennyslentolaivue 35, Vesivehmaa, August 1943. Camouflage colours: upper surfaces Olive Green and Black, under surfaces DN-colour, standard Eastern Front markings Yellow, serial Black, tail stripes White.

FR-139 of T-LeLv 35 is pushed by manpower away from the landing area at Vesivehmaa in February 1943. The rudder shows one style of identification markings, the bars denoting a section of UK 13, which was one course for reserve officers. (Aarno Juurinen)

Fokker D.XXI, FR-156, Täydennyslentolaivue 35, Kauhava, August 1944. Camouflage colours: upper surfaces Olive Green and Black, under surfaces DN-colour, standard Eastern Front markings Yellow, serial Black, tail number White.

FR-156 of T-LeLv 35 at Kauhava, where the unit arrived on 22 June 1944. The second flight possessed at this point more than ten Fokkers, using the tactical number 0 instead of 10. (Lars Bergman)

Fokker D.XXI, FR-156, Täydennyslentolaivue 35, Kauhava, August 1944.

Fokker C.X

Fokker C.X FK-107 at Utti shortly after delivery to LLv 14, which took place on 2 August 1941. Before that it was used by T-LLv 17 in the advanced pilot training role. Quite exceptionally the wing tip undersides are missing the Yellow colour. (Lassi Eskola collection)

Purchase

A natural development of the Dutch Fokker C.V reconnaissance and light bomber aircraft was the C.X. Production consisted of 39 Dutch and 36 licence manufactured aircraft, making the total of 75 airframes. The Finnish Air Force opted for Fokker's new design and four aircraft were bought from Holland, arriving in January 1937, with serials FK-78–FK-81.

The acquired licence for local production yielded 30 aircraft, which were manufactured in 1938 with serials FK-82–FK-111. In 1942 the final batch of five aircraft was assembled, with serials FK-111–FK-115, the first one being issued for the second time. The Finnish Air Force had a total of 39 Fokker C.X aircraft.

Employment

The main users of the C.X were the army co-op squadrons *LLv* 10, 12, 14 and 16 through all three wars in 1939–1944: Winter War, Continuation War and Lapland War.

Only two C.Xs (FK-79 and FK-107) were assigned to training duties, acting as advanced pilot trainers for two-seat reconnaissance and army co-op aircraft. The main user was *T-LLv* 17, which received FK-107 on 19 June 1941. FK-97 arrived on 30 July 1941, releasing FK-107 to combat squadrons. The former's career extended to 23 December 1941, when it was flown to the depot for overhaul.

Colours

FFK-79 of T-LLv 17 at Pori in September 1941. This plane was different in having a Bristol Perseus engine instead of a Bristol Pegasus. Also the Ratier propeller was a non-standard item. (Author's collection)

The four Dutch-built aircraft (FK-78–FK-81) were painted in local colours with Dark Brown Kaki upper sides and aluminium lacquer lower sides. Two of these (FK-78 and 79) were re-painted before the Winter War into Finnish Olive Green upper sides with Light Grey lower side colouring.

The first two batches of licence-produced aircraft (FK-82–FK-111) in 1938 received the contemporary standard camouflage of Olive Green upper sides and Light Grey lower sides. Since only two C.Xs served in the training role, they both carried this camouflage through that employment.

*Fokker C.X, FK-107, Täydennyslentolaivue 17, Karvia, July 1941.
Camouflage colours: upper surfaces Olive Green and under surfaces
Light Grey, non-standard Eastern Front markings Yellow, serial Black.*

*Fokker C.X FK-107 of 3/LLv 14 preparing for a take-off at Utti in early August 1941. During the mobilization for
the Continuation War it was handed over to training unit T-LLv 17 on 19 June 1941. (Lassi Eskola collection)*

FIAT G.50

Purchase

The outbreak of the Second World War speeded up purchase of all war material, and on 23 October 1939 a sales contract with FIAT for 25 G.50 fighters was signed. Another ten planes were added after the outbreak of the Winter War.

The first plane arrived in Finland on 18 December 1939 and was wrongly coded as SA-1. The second Fiat was received on 2 January 1940 and coded SA-2. These codes were changed to FA-1 and FA-2 by an amendment of 26 January 1940.

All planes were planned to travel by rail through Germany. The first two did this but the six following aircraft were caught at a Baltic Sea port and sent to Switzerland. All the others were stopped at the German border. Those in Switzerland were transported first to Amsterdam and later to Sweden by sea. The rest were shipped from Leghorn to Gothenburg, Sweden. The fighters were assembled there and then flown to Finland. Serials FA-3–FA-35 were given to these aircraft.

FIAT FA-25 of HLeLv 30 at Utti in March 1944. It has the full Warpaint of Olive Green and Black tops and DN-colour bottoms with subdued insignia. The latter was introduced on 12 January 1944, to be applied at the next repair or major overhaul. (Anssi Hartiala)

FA-31 of HLeLv 30 stalled on landing at Utti on 24 March 1944 and ended up on its nose. This unit used the FIATs as an intermediary type to train pilots to fly the Messerschmitt Bf 109G. (Anssi Hartiala)

The first arrived on 11 February and all but one by 12 March. The last plane was received on 19 June 1940. Although two were lost en route, thirty-three of the 35 aircraft ordered arrived in Finland.

Employment

The first user was fighter squadron *LLv* 26, operating the FIATs for the last weeks of the Winter War. Service continued almost through the Continuation War. In February and March 1944 a dozen FIATs were handed over to *HLeLv* 30, to act as an intermediary type in training for the Messerschmitt Bf 109 G.

On 1 and 2 June 1944 seventeen of the remaining FIATs were delivered to *T-LeLv* 35, which was the advanced training squadron for fighter and reconnaissance pilots. After the cessation of hostilities on 4 September 1944 the FIATs were grounded two weeks later. *T-LeLv* 35 was disbanded on 27 November 1944 and the FIATs transferred to LeSK.

Flying was resumed in early August 1945 and *LeSK* operated the FIATs until FA-10 flew for the last time on 13 December 1946.

Colours

By the time the FIATs entered service with *T-LeLv* 35, they all had gone through either repair or a major overhaul at the State Aircraft Factory. There all had received the 30 September 1940 regulation Warpaint of Olive Green and Black upper surfaces with DN-colour lower surfaces, keeping these colours to the end of their flying career.

FA-11 at Immola on 28 June 1944, on the way to Kauhava to T-LeLv 35. The standard Warpaint was applied at the factory on 13 August 1943. Though the machine looks pristine here, the sturdy airframe has had four and a half years of hard use and is approaching the end of its life. (PK-Photo)

FIAT G.50 serialled FA-19 belonging to HLeLv 30 ready for a training mission from Utti in April 1944. After their four years of service as an interceptor, they ended the flying career as advanced trainers.

FA-25 of T-LeLv 35 at Kauhava in August 1944. The upper half of the nose has been over-painted in camouflage colours, as stipulated on 13 June 1944. Behind is FA-11 similarly treated. (Lars Bergman)

The pilot of FA-28 (left) of T-LeLv 35 lost control while taxiing at Kauhava on 18 August 1944 and collided with FA-2. This advanced training squadron operated a total of eighteen FIATs, but only for three months before the war ended. (National Archives)

Serials

The FIATs wore the standard serials as specified on 20 March 1934, the height of the serial was 2½ times of the arm of the fuselage swastika.

Since the fighters were assembled in Sweden, the serial font was of standard Swedish style. After factory repairs of overhauls the style complied with the Finnish standard SFS Z.I.1.

S/n	C/n	Delivered	Struck off charge	Remarks	Hours
FA-2	4740	30 May 1944	4 May 1946	W/o Rissala 7 Feb 1946	
FA-9	4737	2 Jun 1944	26 Sep 1944	W/o Utti 2 Jun 1944	
FA-10	3609	1 Jun 1944	2 Jan 1950	Last flight 13 Dec 1946	
FA-11	4733	28 Jun 1944	2 Jan 1950	Last flight 30 Sep 1946	
FA-16	3613	2 Jun 1944	2 Jan 1950	Last flight 22 Oct 1946	
FA-18	3601	2 Jun 1944	2 Jan 1950	Last flight 9 Dec 1946	
FA-19	4725	2 Jun 1944	13 Oct 1944	W/o Kauhava 14 Jun 1944	
FA-22	4646	1 Jun 1944	31 May 1945	Last flight 28 Aug 1944	
FA-25	3614	2 Jun 1944	22 Nov 1945	W/o Kauhava 18 Oct 1945	
FA-26	4743	2 Jun 1944	31 May 1945	Into storage 25 Feb 1945	
FA-27	4944	2 Jun 1944	2 Jan 1949	W/o Kauhava 16 Sep 1944	
FA-28	4731	2 Jun 1944	31 May 1946	Last flight 14 Sep 1944	
FA-29	3606	2 Jun 1944	31 May 1946	Last flight 19 Sep 1944	
FA-31	4939	2 Jun 1944	2 Jan 1950	W/o Kauhava 26 Sep 1946	
FA-32	4726	1 Jun 1944	1 Dec 1945	Worn out	
FA-33	4745	1 Jun 1944	31 May 1946	Last flight 19 Sep 1944	
FA-34	4734	2 Jun 1944	29 Sep 1945	Into storage 9 Nov 1944	
FA-35	4943	2 Jun 1944	2 Jan 1950	Last flight 22 Oct 1946	

This table contains only those FIAT G.50s, which flew in the advanced training role, under given period.

Behind a Pyry is FA-18 of LeSK parked on the ramp at Kauhava in April 1946. It still wears the Warpaint applied on 12 July 1942, with occasional patch-ups and over-painting of the Yellow Eastern Front markings. (Erkki Jaakkola)

FIAT G.50, FA-2, Täydennyslentolaivue 35, Kauhava, August 1944. Camouflage colours: upper surfaces Olive Green and Black, under surfaces DN-colour, standard Eastern Front markings Yellow, serial Black.

FA-2 of T-LeLv 35 parked on the ramp at Kauhava in August 1944. The Warpaint was applied at the field air depot on 13 January 1944, meeting all contemporary regulations. (Lars Bergman)

FIAT G.50, FA-25, Täydennyslentolaivue 35, Kauhava, March 1944. Camouflage colours: upper surfaces Olive Green and Black, under surfaces DN-colour, standard Eastern Front markings Yellow, serial Black.

FA-25 on 20 February 1941. The upper and lower colour demarcation line runs exceptionally down at the corner of the fuselage. The camouflage looks like Warpaint, but is actually a few patch-ups over the original made on FA-25 of HLeLv 30 at Utti in April and May 1944. (Anssi Hartiala)

FA-10 of LeSK is refuelled traditionally by hand pump at Kauhava in summer 1946. The last flight of the type was made by this particular plane on 13 December 1946. (Finnish Air Force Museum)

FA-18 of LeSK on the platform at Kauhava in April 1946. Only the missing yellow Eastern Front markings and roundels, introduced on 1 April 1945, made a difference from the wartime appearance. The victory markings on the fin were applied after the hostilities, but actually show the score of this plane. (Erkki Jaakkola)

Twin-Engine Trainers

- Avro Anson I
- Ilyushin DB-3M
- Hanriot H.232

- Bristol Blenheim I
- Airspeed AS.6E Envoy
- Tupolev USB

Avro Anson I

Purchase

The Anson was a British twin-engined monoplane designed originally in 1934 for coastal surveillance duties. It was developed into a trainer for multi-engine plane aircrews and built in ten versions in large numbers, a total of 11,020 examples.

For Bristol Blenheim bomber aircrew training, Finland bought from A.V. Roe three Ansons, of which the first arrived on 1 October 1936 and the last on 20 January 1937. The serials were at first AN 101–AN 103, until changed to the Finnish standard AN-101–AN-103 in the next major overhaul or repair.

Anson coded AN 101 after arrival at Immola on 23 October 1936. It was assigned as a bomber aircrew trainer to LAs 6 (Air Station 6), which operated two bomber squadrons, LLv 44 and 46, soon to be equipped with Blenheims. (Finnish Air Force)

AN 102 of LAs 6 at Luonetjärvi in March 1937. The camouflage is regulation Olive Green upper with Light Grey lower sides. The significance of the two silver bars on the rudder has remained a mystery. (Finnish Air Force)

AN 103 of LAs 6 at the factory at Tampere in August 1937, repaired from minor landing damage. This plane was the first one to be lost, in a night take-off on 26 February 1940. (Author's collection)

Employment

The Ansons were delivered immediately after arrival to *LAs* 6 (Air Station 6), which operated bomber squadrons *LLv* 44 and *LLv* 46. On 1 January 1938 this air station was re-named *LentoR* 4 (Aviation Regiment 4), where the three Ansons continued their service as aircrew trainers.

In the mobilization of the Winter War in October 1939, a supplementary regiment *T-LentoR* 4 was established within *LentoR* 4, taking care of the training duties.

AN-103 was lost on 26 February 1940 in a crash and the other two Ansons returned to *LentoR* 4, after *T-LentoR* 4 was reduced first on 29 March 1940 to a squadron, *T-LLv* 47, which was disbanded on 20 July 1940. AN-102 flew with *LeR* 4 until a crash on 3 March 1943.

AN-101 was handed over to the air force HQ in July 1942, where it was used to the end of 1946. It made the last Anson flight on 1 July 1947.

AN 101 of T-LentoR *4 taxis for take-off at Luonetjärvi on 7 March 1940. This was a temporary training outfit within* LentoR *4, the bomber regiment. The Avro factory finish is still in good shape, after three years of service. (SA-kuva)*

AN-101 of IlmavE as a VIP transport. It is seen above at Helsinki Malmi in November 1942. During damage repair at the factory the earlier polished aluminium cowlings were also camouflaged. (Author's collection)

Colours

The factory finish for the Ansons was the regular Finnish Air Force camouflage of Olive Green upper and Light Grey lower surfaces, which was specified for planes of wooden and fabric skinning. These colours existed on the Ansons as long as they flew.

AN-101 of the air force HQ, which received this plane on 19 July 1942 as a VIP transport. It came from the factory after damage repair and was finished in Finnish colours with Light Grey under surfaces. The serial was also altered to Finnish specifications. (Author's collection)

AN-102 trainer of LeLv 48 at Luonetjärvi in August 1942. The cowlings have been painted in a recent overhaul, either fresh Olive Green or even Black. The camouflage is regulation solid Olive Green with Light Grey undersides. (Lassi Eskola collection)

Serials

Upon arrival the Anson bore English style serials, which were applied without the dash.

In a repair or major overhaul at the factory, Finnish style serials, as specified on 20 March 1934, were applied. The height of the serial was 2½ times of the arm of the fuselage swastika. The style complied with the Finnish standard SFS Z.I.1.

AN-102 of T-LeLv 17 at Luonetjärvi, where it crashed on 3 March 1943. The Anson had the facilities to train pilots, navigators, wireless operators and gunners. (Authors collection)

S/n	C/n	Delivered	Struck off charge	Remarks	Hours
AN-101		6 Oct 1936	1 Oct 1952	Last flight 3 Jul 1947	1243
AN-102		28 Nov 1936	13 Jul 1943	W/o Luonetjärvi 3 Mar 1943	1531.40
AN-103		28 Jan 1937	11 Nov 1940	W/o Rovaniemi 26 Feb 1940	525.30

Avro Anson I, AN 101, Lentoasema 6, Immola, October 1936. Camouflage colours: upper surfaces Olive Green, under surfaces Light Grey, serial Black.

AN 101 arrived at Immola on 23 October 1936. Immola became three weeks later the official base for the bomber station LAs 6, which possessed two squadrons, LLv 44 and 46. (Finnish Air Force)

Avro Anson I, AN-102, Lentolaivue 48, Luonetjärvi, July 1942. Camouflage colours: upper surfaces Olive Green, under surfaces Light Grey, engine cowling aluminium dope, standard Eastern Front markings Yellow, serial Black.

AN-102

Anson AN-102 of LeLv 48 at Luonetjärvi in July 1942, still having polished aluminium cowlings. Until the establishment of T-LeLv 17 on 28 November 1942, LeLv 48 was in charge of the bomber aircrew supplementary training. (Author's collection)

Ilyushin DB-3M

Purchase

During the Winter War in 1939–40 a number of DB-3M bombers were captured. Five of them were refurbished by the State Aircraft Factory and given Finnish serials VP-101, 102 etc. according to the 12 February 1940 order. The first one became VP-101 and while the others were being completed, the serials were amended in December 1940 to VP-11–VP-15.

After the outbreak of the Continuation War six more aircraft were bought from the German war booty depots. These arrived in Finland on 12 September 1941 and they were serialled DB-16–DB-21. The Finnish Air Force had thus a total of eleven DB-3M bombers in its inventory.

Employment

At the outbreak of the war on 25 June 1941 bomber squadron *LLv* 46 had four DB-3M planes, which were used only for training purposes. Then they were delivered to the new LLv 48, which was established at Luonetjärvi on 23 November 1941, also under *Lentorykmentti* 4.

Their task was the advanced training of personnel for the regiment's needs. Normally only one DB-3 was available for bomber crew training purposes. In June 1942 the Ilyushin DB-3M was classified as a bomber and the training career of the type was over.

Colours

DB-3M serialled VP-14 after refurbishing at the State Aircraft Factory at Tampere. On 14 March 1941 it was handed over to LLv 46, which used it only in training. It wears the Warpaint of Olive Green and Black with aluminium dope undersides. The latter was the regulation finish for metal skinned planes. (Aaretti Nieminen)

The first DB-3M, which was captured on 29 January 1940 in fully airworthy condition, retained its aluminium dope overall finish. The Soviet markings were painted over and the Finnish national insignia and serial number VP-101 were applied instead.

VP-11 of LLv 46 at Luonetjärvi shortly before it was lost on 30 June 1941 in a ditching. All markings are according to the regulations. The Yellow theatre markings, fuselage band and lower wing tips, were applied on 18 June 1941. (Lassi Eskola collection)

VP-12 of LLv 46 on a visit to Laajalahti on 5 July 1941. This plane suffered a landing accident a month later and was sent to the factory for repairs. The aluminium dope finish on the lower surfaces glitters well on the engines. (Esa Laukkanen)

The serial was changed to VP-11 in December 1940 and in this connection the aircraft received the new Warpaint introduced on 30 September 1940: Olive Green and Black upper sides, lower sides retaining the aluminium dope finish.

The other four war booty DB-3Ms received similar Warpaint during major repair and overhaul at the State Aircraft factory, with serials VP-12–VP-15.

Serials

On 17 September 1941 the rather confusing serial numbering of captured bombers was solved by giving separate type identifications, for the DB-3M this was DB with the consecutive numbers being the same, here just DB-12 and 15 since VP-11 and 14 were lost in flying accidents and VP-13 sent to Germany.

The serials were as specified on 20 March 1934, the height of the serial was 2½ times of the arm of the fuselage swastika. The style complied with the Finnish standard SFS Z.I.1.

S/n	Delivered	Struck off charge	Remarks	Hours
VP-11	28 Mar 1940	20 May 1942	W/o Hirvaslampi 30 Jun 1941	82.15
VP-12	24 Jun 1941		Damaged 4 Aug 1941	
VP-13	21 Feb 1941		To Germany 12 May 1941	
VP-14	14 Mar 1941	17 Jul 1941	W/o Konnevesi 2 Jul 1941	26.05
VP-15, DB-15	24 Jun 1941		Bomber with LeLv 48 1 Jun 1942	

This table contains only those Ilyushin DB-3Ms which flew in the training role during the given period.

DB-15 trainer of LLv 48 at Luonetjärvi at the beginning of 1942. The serial was amended into this form on 17 September 1941. The training career of the type ended by June 1942, when all DB-3Ms of the unit were classified as bombers. (Lassi Eskola collection)

Ilyushin DB-3M, VP-11, Lentolaivue 46, Luonetjärvi, June 1941. Camouflage colours: upper surfaces Olive Green and Black, under surfaces aluminium dope, standard Eastern Front markings Yellow. Serial Black on Olive Green paint and Olive Green on Black paint.

VP-11 trainer of LLv 46 at Luonetjärvi, after the addition of the Yellow Eastern Front markings on 18 June 1941, a week before the Continuation War began, in the Russian air attacks on Finnish locations. (Lassi Eskola collection).

Left: VP-11 of LLv 46, while taxing on the soft airfield at Rissala, ended up on its nose on 7 May 1941, merely bending the propellers a little. It has the regulation Warpaint with aluminium dope undersides. (Author's collection)

Ilyushin DB-3M of Lentolaivue 48, DB-15, Luonetjärvi, February 1942. Camouflage colours: upper surfaces Olive Green and Black, under surfaces aluminium dope, standard Eastern Front markings Yellow. Serial Black on Olive Green paint and Olive Green on Black paint.

DB-15 trainer of LLv 48 at Luonetjärvi in February 1942. This was for months the only actual plane which the unit had available for aircrew training. The colours and markings are very much standard. (Author's collection)

Hanriot H.232

Purchase

The H.232 was a French twin-engined trainer design from 1937. Thirty-seven aircraft were built before the German occupation in June 1940. Twenty-six trainers ended up in German hands. Three planes were sold to the Finnish Air Force. They were flown to Finland, two arriving on 24 July 1941 and one lost en route. The serials were HT-191–193.

Employment

After refurbishing by the State Aircraft Factory the Hanriots were handed over on 21 August 1941 to *T-LLv* 17, which was an advanced training squadron for bomber and reconnaissance aircrews. It was disbanded on 1 November 1941 and *LLv* 46 took over the training. From 27 November 1941 *LLv* 48 carried on the training, until *T-LeLv* 17 was re-established on 15 December 1942.

HT-191 left the trainer service after an accident on 28 August 1942 and HT-193 retired on 17 February 1945, following in a bad landing.

HT-191 of T-LLv 17 *taxied into a ditch at Karvia on 28 September 1941. The plane was about to be handed over to LLv 46, but it was delayed by a few days. The markings conform with all contemporary regulations. (Finnish Air Force)*

HT-191 of T-LLv 17 at Pori, where this plane arrived on 21 August 1941. It was refurbished by the State Aircraft Factory, which applied the regular trainer colours of solid Olive Green upper and Light Grey lower surfaces. (Finnish Air Force)

Colours

The Hanriots were flown to Finland in French camouflage, but it did not exist long as the State Aircraft Factory refurbished and re-painted the two Hanriots in regulation trainer colours of Olive Green tops and Light Grey undersides, which remained on them until the end of their flying career.

Serials

The transfer serials were off-standard, but the factory applied the standard serials as specified on 20 March 1934. The height of the serial was 2½ times of the arm of the fuselage swastika. The style complied with the Finnish standard SFS Z.I.1.

S/n	C/n	Delivered	Struck off charge	Remarks	Hours
HT-191	21	21 Aug 1941	2 Jan 1950	Last flight 16 Jan 1945	290.15
HT-192	23			W/o Brandenburg 19 Jul 1941	
HT-193	24	21 Aug 1941	2 Jan 1950	W/o Tampere 15 Feb 1945	146.55

HT-191 of LLv 48 photographed at Luonetjärvi in February 1942. This squadron was in charge of the bomber aircrew supplementary training up to 28 November 1942, when T-LeLv 17 was re-established. (Finnish Air Force)

Hanriot H.232, HT-191, Lentolaivue 48, Luonetjärvi, February 1942. Camouflage colours: upper surfaces Olive Green, under surfaces Light Grey, standard Eastern Front markings Yellow, serial Black.

HT-191 of LLv 48 parked at Luonetjärvi in February 1942. This type was used exclusively in supplementary pilot training for the bomber crews. (Finnish Air Force)

Hanriot H.232, HT-193, Täydennyslentolaivue 17, Pori, September 1941. Camouflage colours: upper surfaces Olive Green, under surfaces Light Grey, standard Eastern Front markings Yellow, serial Black.

HT-193 of T-LLv 17 pictured at Pori, where the plane arrived on 21 August 1941. Apart from the serial, this plane can be identified by the cranked pitot on the nose. (Author's collection)

Bristol Blenheim I

Purchase

The Blenheim was a British bomber design from 1934 and it was built in six major versions a total of 6198 examples. The Finnish Ministry of Defence ordered 18 Blenheim Is (BL-104–BL-121, I series) from the Bristol Co. as the first foreign customer on 6 October 1936. The first two planes arrived in Helsinki on 29 July 1937 the last two being delivered on 27 July 1938.

On 12 April 1938 a production license was acquired, and 15 Blenheim IIs were subsequently ordered from the State Aircraft Factory (VL) on 6 April 1939 (series II). Before production had started World War II and the Finnish-Soviet Winter War broke out and two batches were ordered from England.

The first Blenheim to arrive at LLv 48 was this BL-151, coming on 5 February 1942 to Luonetjärvi. The stay lasted only a month, when BL-135 replaced it. BL-151 has the regulation Warpaint of Olive Green and Black with aluminium dope undersides. (Lassi Eskola collection)

BL-135 was the second to arrive with LLv 48 based at Luonetjärvi, on 6 March 1942. This particular plane already had dual controls, which was much better suited for training purposes. Colours are fully by the contemporary orders. (Author's collection)

BL-140 of LLv 48 had a tyre burst and went on its nose at Luonetjärvi on 27 April 1942. The camouflage pattern can be seen on the wings. It was basically the same for all types. (Lassi Eskola collection)

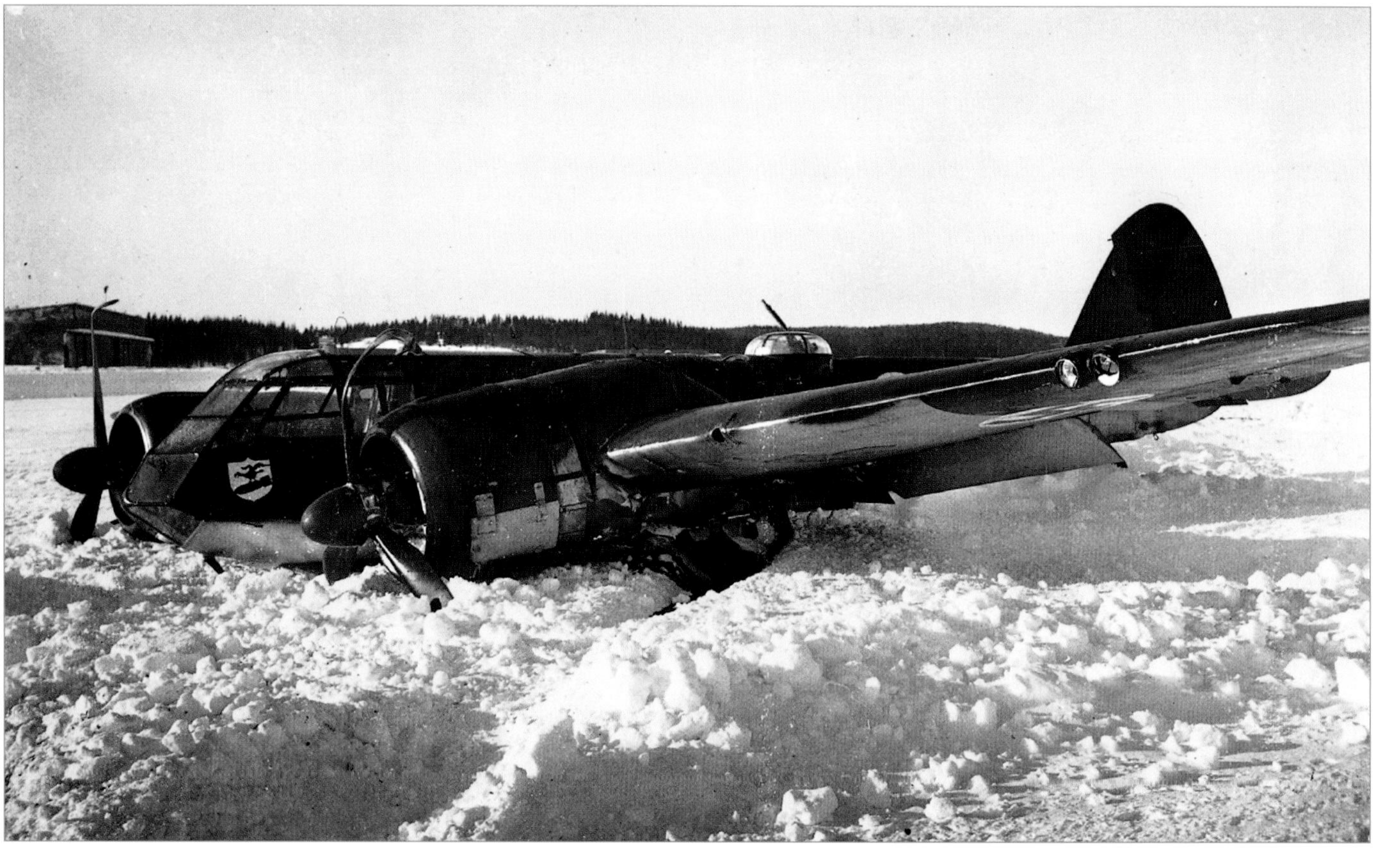

BL-111 of T-LeLv *17 bellied at Luonetjärvi on 22 February 1943. This new advanced training squadron took charge of all bomber crew training. The nose emblem belongs to the earllier user 3/*LeLv *44. (Author's collection)*

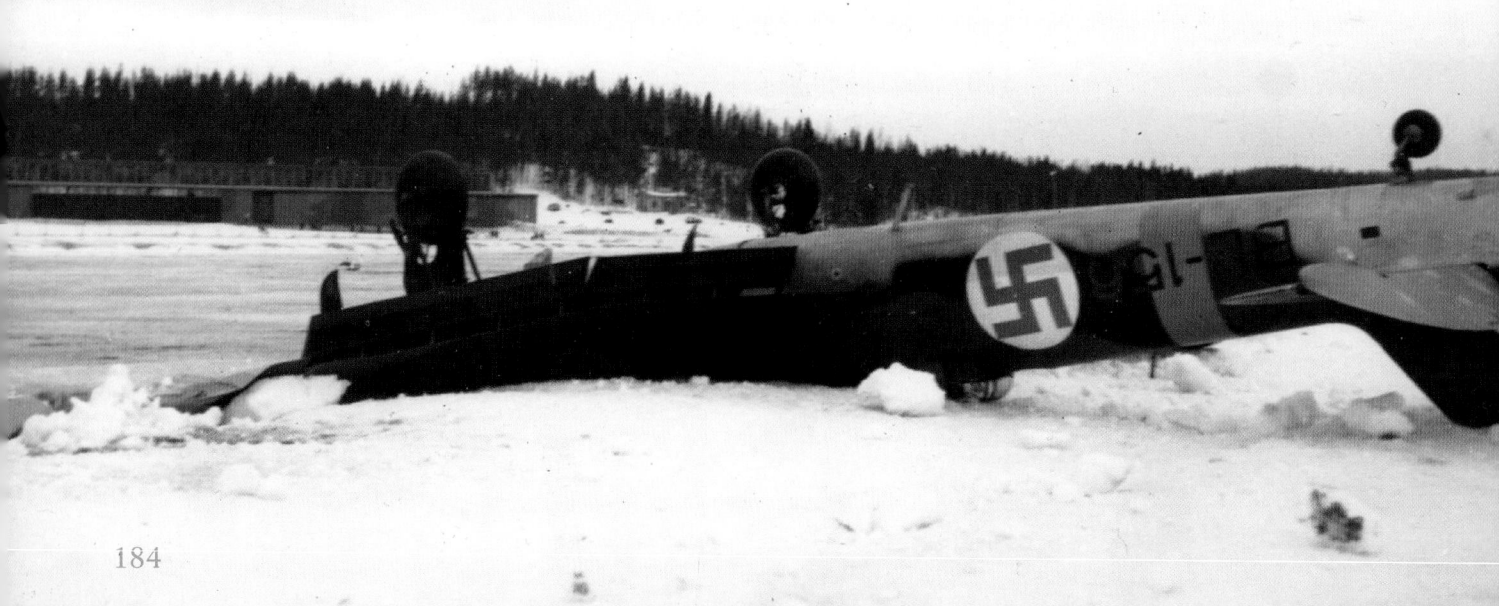

Twelve "long-nose" Blenheim IVs (BL-122–BL-133; series III) were handed over to Finnish crews on 17 January 1940. Ten aircraft arrived in Finland four days later (one aircraft disappeared over the North Sea and the other arrived in Finland on 5 June 1940, after an accident in Sweden).

Series IV consisted of 12 Blenheim Is (BL-134–BL-145) which arrived in Finland on 26 February 1940 flown by British transfer crews.

During the short peace period the Aircraft Factory started production of the previously ordered series II (BL-146 - BL-160), the first plane being delivered on 14 June 1941 and the last on 9 January 1942.

On 7 January 1942 an order was placed at the factory for 30 Blenheim IIs (series V, coded BL-161–BL-190, delivered from 28 July to 26 November 1943) and for 10 Blenheim IVs (series VI, coded BL-196–BL-205, delivered from 26 February to 15 April 1944).

VL thus produced a total of 45 Blenheim IIs and 10 Blenheim IVs. The Finnish Air Force had a total of 97 Blenheims.

BL-156 of T-LeLv *17 flipped over at Luonetjärvi on 16 March 1943. The lower surfaces of the Warpaint are still finished in aluminium dope. The repair took eight months and then the service comntinued in bomber squadron* LeLv *42. (Author's collection)*

BL-115 of T-LeLv 17 in the landing run at Luonetjärvi on 4 April 1944. The Warpaint with DN-colour undersides was applied during a damage repair at the factory on 30 August 1943. (SA-kuva)

BL-111 of T-LeLv 17 landing at Luonetjärvi on 4 April 1944. Its Warpaint with DN-colour undersides was also factory applied on 17 August 1943. (SA-kuva)

When BL-111 of T-LeLv 17 landed to Luonetjärvi on 11 August 1944, the landing gear went in and the plane bellied. The repair took nine weeks. (Kustaa Kotsalo)

Employment

The vast majority of the Blenheims served as bombers in three wars: Winter War, Continuation War and Lapland War. Every one of the four bomber squadrons was equipped with the Blenheim at one time.

The Blenheim entered training duties on 5 February 1942, when BL-151 was transferred to *LLv* 48, then in charge of the bomber aircrew advanced training. Next March two more Blenheims arrived. These had been converted to trainers by installing dual controls.

On 28 November 1942 a dedicated advanced training squadron, *T-LeLv* 17, was re-established at *LeR* 4. All training personnel and equipment held by *LeLv* 48 were transferred to *T-LeLv* 17.

More Blenheims kept coming and the unit had an average of 8–10 Blenheims serviceable until the end of the hostilities. The Continuation War ended in an armistice on 4 September 1944 and a week later *T-LeLv* 17 was disbanded.

Colours

BL-111 of T-LeLv 17 at Luonetjärvi on 15 April 1944. This plane is from the seven year old I series, The bomb bays were open but the racks themselves were removed. It also had dual controls. (SA-kuva)

When transferred to training units the Blenheims wore the standard camouflage introduced on 30 September 1940: upper surfaces in Warpaint of Olive Green and Black with lower surfaces in aluminium dope, which was the finish for metal covered aircraft. These upper side colours remained so to the end of the hostilities and beyond.

On 7 May 1942 an order was given to paint the under surfaces of the warplanes in Light Blue-Grey DN-colour (*RLM Hellblau* 65), at the next repair or overhaul at the factory. Out of the dozen Blenheims serving with *T-LeLv* 17, only six had the DN-colour under sides, these being BL-106, 111, 115, 156, 179 and 180.

S/n	C/n	Delivered	Struck off charge	Remarks	Hours
BL-106		5 Apr 1944		Damage at Kauhava 8 Jun 1944	
BL-109		21 Sep 1943		To factory 25 Jul 1944	
BL-111		17 Feb 1943		To *PLeLv* 41 4 Dec 1944	
BL-115		27 Sep 1943		Damage at Luonetjärvi 9 Sep 1944	
BL-120		27 Feb 1943		Damage at Luonetjärvi 29 Feb 1944	
BL-135		6 Mar 1942		W/l Luonetjärvi 9 Jul 1943	
BL-138		30 Apr 1943		W/o Luonetjärvi 9 Mar 1944	
BL-140		13 Mar 1942		W/o Luonetjärvi 18 Aug 1943	
BL-151		5 Feb 1942		To *LLv* 42 6 Mar 1942	
BL-156		20 Feb 1943		Damage at Luonetjärvi 16 Mar 1943	
BL-160		17 Apr 1943		To *LeLv 42* 2 May 1943	
BL-179		13 Nov 1943		To *PLeLv* 45 4 Dec 1944	
BL-180		10 Nov 1943		W/o Laukaa 11 Oct 1944	

This table shows only those Blenheims which served in training duties during the given period.

BL-106 of T-LeLv 17 flipped over in a landing at Kauhava on 8 June 1944. It has a full Warpaint with DN-colour undersides applied at the factory on 13 March 1943. (Author's collection)

Blenheim BL-180 of T-LeLv 17 at Luonetjärvi just before the end of the Continuation War, which occurred on 4 September 1944. This and BL-179 were the only ones, which came as new directly from the State Aircraft Factory to T-LeLv 17. (Kustaa Kotsalo)

Bristol Blenheim I, BL-135, Lentolaivue 48, Luonetjärvi, June 1942. Camouflage colours: upper surfaces Olive Green and Black, under surfaces aluminium dope, standard Eastern Front markings Yellow. Serial Black on Olive Green paint and Olive Green on Black paint.

BL-135 of LeLv 48 at Kauhava at the end of May/ early June 1942. It flew radio calibration flights. This was an ex-British machine, which received the Warpaint at the factory on 26 February 1942. (Author's collection)

Bristol Blenheim II, BL-180, Täydennyslentolaivue 17, Luonetjärvi, June 1944. Camouflage colours: upper surfaces Olive Green and Black, under surfaces DN-colour, standard Eastern Front markings Yellow. Serial Black on Olive Green paint and Olive Green on Black paint.

BL-180 of T-Lelv 17 at Luonetjärvi in June 1944. This was converted to a trainer with dual controls whilst under construction at the factory. The Warpaint was applied on 20 October 1943. (Author's collection)

Airspeed AS.6E Envoy

Purchase

On 8 November 1941 German fighters shot down in error a Finnish airliner, a D.H. 89 Dragon Rapide, registered as OH-BLB and named *"Lappi"*. To replace this the Germans delivered one Envoy, which they had earlier seized from Czechoslovak airline *CZA*. This plane was delivered to Insterburg, Germany on 22 January 1942 and flown to Finland.

Employment

After arrival it was refurbished by the air depot and given serial EV-1. On 28 April 1942 it was handed over to *LLv* 48 for bomber aircrew training. It was delivered to the new *T-LeLv* 17 on 15 December 1942. On 1 June 1943 the air force Signals School took over the plane, until it crashed on 31 July 1943, having logged 148 hours in Finnish service.

Colours

The aircraft was finished in overall aluminium dope. Aside from the national insignia and proportionally too large serial numbers, the only colour added was the Eastern Front yellow under the wing tips and later the rear fuselage band.

Serials

The Envoy wore a serial which resembled the one coming to effect on 20 March 1934, but it was larger than specified and with a slightly different font.

Envoy serialled EV-1 arrived at LLv 48, based at Luonetjärvi, on 25 April 1942. The yellow colour under the wing tips is a remnant from the earlier German ownership. (Lassi Eskola collection)

EV-1 of LeLv 48 at Luonetjärvi in May 1942. This was a replacement aircraft for a mistakenly shot down Finnish Dragon Rapide airliner in the previous November. (Lassi Eskola collection)

Airspeed A.S.6E Envoy, EV-1, Täydennyslentolaivue 17, Luonetjärvi, May 1943. Aluminium dope finish overall, standard Eastern Front markings Yellow, serial Black.

EV-1 of T-LeLv 17 in front of Luonetjärvi hangar in May 1943. The next user was IlmavVK (air force Signals School) from 1 June 1943. (Lassi Eskola collection)

Tupolev SB

Purchase

The SB was a Soviet fast bomber design from 1934. It was built in several versions, for a total of 6,992 examples. Several SB bombers were captured in the Winter War. Eight of these were refurbished into flying condition, serials VP-1, 2 etc. These were changed in December 1940 to VP-1–VP-8.

After the outbreak of the Continuation War sixteen SB bombers were bought from the German war booty depots. The aircraft were delivered in three lots packed in crates. The first lot arrived on 5 November 1941 and the last batch came on 27 August 1942 and received serials SB-9–SB-24. A total of twenty-four Tupolev SB aircraft were in the Finnish Air Force inventory.

Since a good number of SBs were purchased, the air force headquarters ordered on 7 September 1942 that SB-6 and SB-8 are to be modified as trainers. All armament was removed, the glass nose received

SB-5 has just come out of a damage repair at the State Aircraft Factory and is seen here at Tampere. It was handed over to T-LeLv 17 as a trainer on 9 February 1943. (Aaretti Nieminen)

SB-8 seen here at Luonetjärvi with T-LeLv 17 in summer 1943. It was converted to a trainer by March 1943, the modification included dual controls, a second cockpit in the observer's position and a nose cone. (Lassi Eskola collection)

Both USBs of T-LeLv 17 *in the hangar at Luonet-järvi during summer 1943. The second cockpit in the observer's position, with separate wind shield, and the nose cone distinguish these from standard bombers. (Lassi Eskola collection)*

a wooden cone, the aircraft was fitted with dual controls and a second cockpit was built in the observer's position. The dorsal position was covered with a sliding hatch. The modifications were carried out on SB-6 by 25 January 1943 and on SB-8 by 1 March 1943. In the Soviet Union this modification was designated as USB.

Employment

The main Tupolev roles in the Finnish Air Force were bomber, maritime reconnaissance and submarine hunting with *LeLv 6*.

The first SB trainer entering service with *T-LeLv 17* was SB-5 on 9 February 1943, but the stay lasted only a little over four weeks. After trainer conversion, SB-6 arrived in February 1943 and SB-8 a month later to *T-LeLv 17*, both flying to the end of the hostilities.

Colours

The Warpaint was introduced on 30 September 1940 and it consisted of Olive Green and Black upper surfaces. Captured aircraft got a splinter pattern camouflage and this was applied on all trainer SBs. Likewise all received the Light Blue-Grey DN-colour undersides, which was introduced on 7 May 1942.

Serials

The Tupolev SBs wore the standard serials as specified on 20 March 1934, the height of the serial was 2½ times of the arm of the fuselage swastika. The style complied with the Finnish standard SFS Z.I.1.

The VP letters in the serial came from *Venäläinen Pommittaja* (Russian Bomber) and all types were marked so, causing some problems in keeping the records. On 17 September 1941 an order was give to change the serial to identify a particular type. The Tupolev SB was given the letters SB to the serial, while the numbers ran as before.

S/n	Delivered	Struck off charge	Remarks	Hours
SB-5	9 Feb 1943		To *LeLv* 6 23 Mar 1943	
SB-6	19 Feb 1943	2 Jan 1950	Into storage 24 Feb 1945	
SB-8	18 Mar 1943	9 Feb 1945	W/o Luonetjärvi 25 Oct 1944	

This table contains only those Tupolev SBs which flew in the training role in the given period.

Tupolev SB, SB-5, Täydennyslentolaivue 17, Luonetjärvi, March 1943. Camouflage colours: upper surfaces Olive Green and Black, under surfaces DN-colour, standard Eastern Front markings Yellow. Serial Black on Olive Green paint and Olive Green on Black paint.

SB-5 of T-LeLv 17 at Luonetjärvi in March 1943. The stay was short and temporary, just waiting for the trainer conversion USBs to arrive. (Author's collection)

Tupolev SB, SB-5, Täydennyslentolaivue 17, Luonetjärvi, March 1943.

Both Tupolev USB trainers of T-LeLv 17, SB-6 and SB-8, at Luonetjärvi in summer 1943. Both have now M-103 engines and the latter still wears the old style cowlings with frontal radiators. (Lassi Eskola collection)

SB-6 of T-LeLv 17 being towed inside the hangar at Luonetjärvi on 30 March 1943. The camouflage and markings cannot be much better shown than here. (SA-kuva)

Tupolev USB, SB-6, Täydennyslentolaivue 17, Luonetjärvi, June 1943. Camouflage colours: upper surfaces Olive Green and Black, under surfaces DN-colour, standard Eastern Front markings Yellow. Serial Black on Olive Green paint and Olive Green on Black paint.

After trainer modification SB-6 arrived at T-1eLv 17, based at Luonetjärvi, on 19 February 1943. It is seen here a few weeks later. (Author's collection)

SB-6 in front of Luonetjärvi hangar in June 1943. It wears full Warpaint, with a splinter pattern designed for captured aircraft only. (Author's collection)

SB-6 of T-LeLv 17 has returned to Luonetjärvi from a training sortie on 30 March 1944. It wears the splinter pattern Warpaint of Olive Green and Black tops and DN-colour bottoms. The splinter pattern camouflage was exclusive for captured aircraft from 7 May 1942 onwards. (SA-kuva)

Gunnery Trainers

- VL Tuisku
- Aero A-32GR
- Blackburn Ripon IIF
- Fokker C.V
- Koolhoven F.K.52

VL *Tuisku*

Purchase

The design of the *Tuisku* (Gust of wind) was started in 1932 by Dipl. Eng Arvo Ylinen, who a while later became the chief designer of the State Aircraft Factory. In spring 1933 the government participated in the costs and ordered one prototype of this multi-purpose trainer biplane. The fuselage was of welded steel tube frame with fabric covering while the wings were of wooden construction covered with fabric. It was powered by a 215 hp Armstrong Siddeley Lynx IV radial engine.

The prototype TU-149 made its maiden flight on 10 January 1934, but was lost nineteen days later during a dive test, due to flutter. The accident was thoroughly investigated and strengthened wings were made to prevent a repeat.

Tuisku prototype TU-149 outside the State Aircraft Factory on the ice at Suomenlinna in January 1934. It wears the new camouflage of Olive Green tops and aluminium dope undersides, which became standard for the next seven years. (VL)
The second Tuisku *prototype was TU-150, seen here at Santahamina in June 1935. All subsequent production machines were similar, except the cowling ring was later deleted. (VL)*

The second prototype TU-150 was ordered on 14 May 1934 and it was first flown on 12 December 1934. After trials the factory received an order for twelve aircraft in February 1935. These aircraft were completed by July 1936 and coded TU-151–TU-162. Additionally TU-163 was built by funds raised by students and inscribed "Pilven Veikko".

Of the I-series eight were land planes, the first four were for observer training and the next four for pilot training, with dual controls. The last four were observer planes fitted with floats.

A sixteen aircraft II-series was ordered on 14 February 1936 and they were completed by July 1937. The serials were TU-164–TU-179. The Finnish sugar company sponsored TU-165, which was inscribed "Sokeri-Sirkku". It was lost on 19 June 1940, but the name was inherited by TU-178.

Of these the first two were originally observer landplanes, the next four next observer floatplanes and the rest pilot training landplanes. After autumn 1939 the *Tuiskus* serves only as landplanes.

The Finnish Air Force had a total of thirty-one *Tuisku* multi-role trainers serving in the pilot, observer and gunnery training roles, plus other flying, bombing, photographing and communication training.

Tuisku TU-153 overshot the landing to Santahamina on 22 June 1936. It was a brand new plane and still performing factory test flights. (VL)

TU-165 was acquired with funds provided by the Finnish sugar company, getting the inscription Sokeri-Sirkku. It is a seen here at Tampere on July 1939. The plane wears the regular colours of Olive Green and aluminum dope. (Author's collection)

Employment

Half of the new *Tuiskus* were placed with the Aviation School (*Ilm.K*) at Kauhava and a couple of individual planes to all land and maritime air stations: *LAs* 1 at Utti, *LAs* 2 at Santahamina, *LAs* 3 at Sortavala, *LAs* 4 at Turkinsaari, *LAs* 5 at Suur-Merijoki and *LAs* 6 at Viipuri. When the regiments were established on 1 January 1938, the Air Fighting School was the main user and individual *Tuiskus* served in all other outfits.

By the Winter War only three aircraft had been lost. *Tuiskus* were concentrated at the Air Fighting School and auxiliary units. Six aircraft were lost between the wars and by the Continuation War *Tuiskus* continued to serve at *LeSK* and advanced training squadrons. After the end of the war five aircraft were put into storage, being worn out.

Sixteen aircraft survived beyond 1944 and they were used for liaison and as hacks by all air force units. After 1948 only three were on duty and the last flight was carried out by TU-151 on 31 January 1950.

Colours

All *Tuiskus* were camouflaged with Olive Green upper and aluminium dope lower sides. This scheme was valid as long as the *Tuisku* flew.

Serials

TU-157 of ISK at Kauhava on 13 August 1939. The large serial was specified for the Tuisku on 13 May 1939. Number 157 of the serial is in Black under the lower wings. They later were white. (Finnish Air Force)

Large and different serials were used until 20 March 1934, when a directive stipulated that the height of the serial was 2½ times of the arm of the fuselage swastika, much smaller than before.

On 13 May 1939 small letters and large numbers in white were ordered to be painted on *Tuiskus*. The numbers were also introduced under both lower wings. This also transferred the national marking from the fuselage to the rudder.

This was done at a major overhaul or repair and was to remain throughout the career of the *Tuisku*.

TU-150 gunnery trainer of T-LLv 17 at Karvia in September 1941. This plane already has White numbers below the lower wings. (Author's collection)

TU-174 of LeSK at Mänkijärvi camp in November 1941. The plane carries small practice bombs on the fuselage rack. The colours and markings are very much by the regulations, Yellow also under both wing tips. (Author's collection)

TU-174 of LeSK at Kauhava in July 1942. The Olive Green paintwork has begun to show wear and tear and a major overhaul will follow in May 1943. (Author's collection)

TU-166 at Kuorevesi after a factory overhaul, shortly before the hand over to LeSK on 19 June 1942. (VL)

Four Tuiskus of LeSK ready for the days exercises at Kauhava on 20 July 1943. The plane from the left are TU-166, 152, 167 and 154. All are in regulation Olive Green tops and alu dope bottoms-(SA-kuva)

A busy scene outside hangar 2 at Kauhava on 20 July 1943. At right is TU-166 of LeSK pushed out to the airfield. (SA-kuva)

TU-152 of LeSK takes off at Laajalahti in July 1943, on the way to the shooting range over the sea. The serial number can be seen under the lower wing. (SA-kuva)

TU-154 of LeSK at Kauhava in March 1943. The camouflage of solid Olive Green and aluminium dope with White serials are fully by the book. (Author's collection)

TU-152 of LeSK is pushed and pulled out to the airfield at Kauhava on 20 July 1943. It is in perfect regulation markings. (SA-kuva)

S/n	C/n	Delivered	Struck off charge	Remarks	Hours
TU-149		10 Jan 1934	29 Jan 1934	W/o Suomenlinna 29 Jan 1934	
TU-150		5 Sep 1935	9 Feb 1945	W/o Oulu 19 Sep 1944	1873.50
TU-151		29 Jun 1936	1 Oct 1952	Last flight 31 Jan 1950	2786.50
TU-152		22 Jun 1936	9 Apr 1949	Into storage 17 Feb 1945	2482.10
TU-153		11 Nov 1936	9 Apr 1949	Into storage 21 Mar 1947	2362.35
TU-154		30 Nov 1936	1 Oct 1952	Into storage 22 Apr 1949	3501.15
TU-155		29 Jul 1936	11 Dec 1952	Into storage 13 Sep 1946	2002.55
TU-156		7 Aug 1936	11 Apr 1949	W/o Utti 11 Feb 1948	2154
TU-157		2 Nov 1936	11 Apr 1949	Into storage 18 May 1948	2321.20
TU-158		17 Aug 1936	30 Apr 1941	W/o Lappajärvi 20 Mar 1941	1159.35
TU-159		27 Jan 1937	14 Mar 1938	W/o Uuras 27 Jan 1938	373.10
TU-160		31 Aug 1936	8 May 1941	W/o Lappajärvi 27 Mar 1941	1054.35
TU-161		31 Aug 1936	11 Apr 1949	Into storage 20 Feb 1945	1868
TU-162		16 Sep 1936	1 Oct 1952	Into storage 9 Nov 1948	2048.55
TU-163		23 May 1937	2 Mar 1946	W/o Simo 7 Jan 1946	1853.35
TU-164		27 May 1937	2 Jan 1950	W/o Tampere 22 Feb 1947	2105.10
TU-165		8 Jun 1937	30 May 1941	W/o Karvia 10 Jun 1940	947.40
TU-166		1 Jun 1937	1 Oct 1952	Into storage 15 Feb 1945	1553.45
TU-167		4 Jun 1937	26 Sep 1944	W/o Onttola 8 Aug 1944	1678
TU-168		29 Sep 1937	16 Dec 1940	W/o Tuusula 16 Nov 1940	729.35
TU-169		30 Jun 1937	1 Oct 1952	Into storage 17 Apr 1948	2235.20
TU-170		21 Jun 1937	16 Aug 1943	W/o Kauhava 7 Jun 1943	1370.30
TU-171		29 Jun 1937	1 Oct 1952	Into storage 10 May 1948	1744.20
TU-172		7 Jul 1937	28 Jan 1938	W/o Kauhava 29 Nov 1937	106.05
TU-173		12 Jul 1937	28 Sep 1940	W/o Pori 23 Aug 1940	784.25
TU-174		22 Jul 1937	1 Oct 1952	Into storage 24 Jul 1947	2189.50
TU-175		23 Jul 1937	1 Oct 1952	Into storage 6 Aug 1947	2166.25
TU-176		3 Aug 1937	1 Oct 1952	Into storage 3 Apr 1947	1965.50
TU-177		28 Jul 1937	5 Mar 1941	W/o Jalasjärvi 4 Feb 1941	547
TU-178		29 Jul 1937	2 Jul 1953	Into storage 22 Jul 1948	1808.15
TU-179		21 Sep 1937	1 Oct 1952	Into storage 13 Nov 1947	1898.25

TU-167 of LeSK in flight over Laajalahti camp area on 20 July 1943. The school's gunnery practices were carried out here over the sea. (SA-kuva)

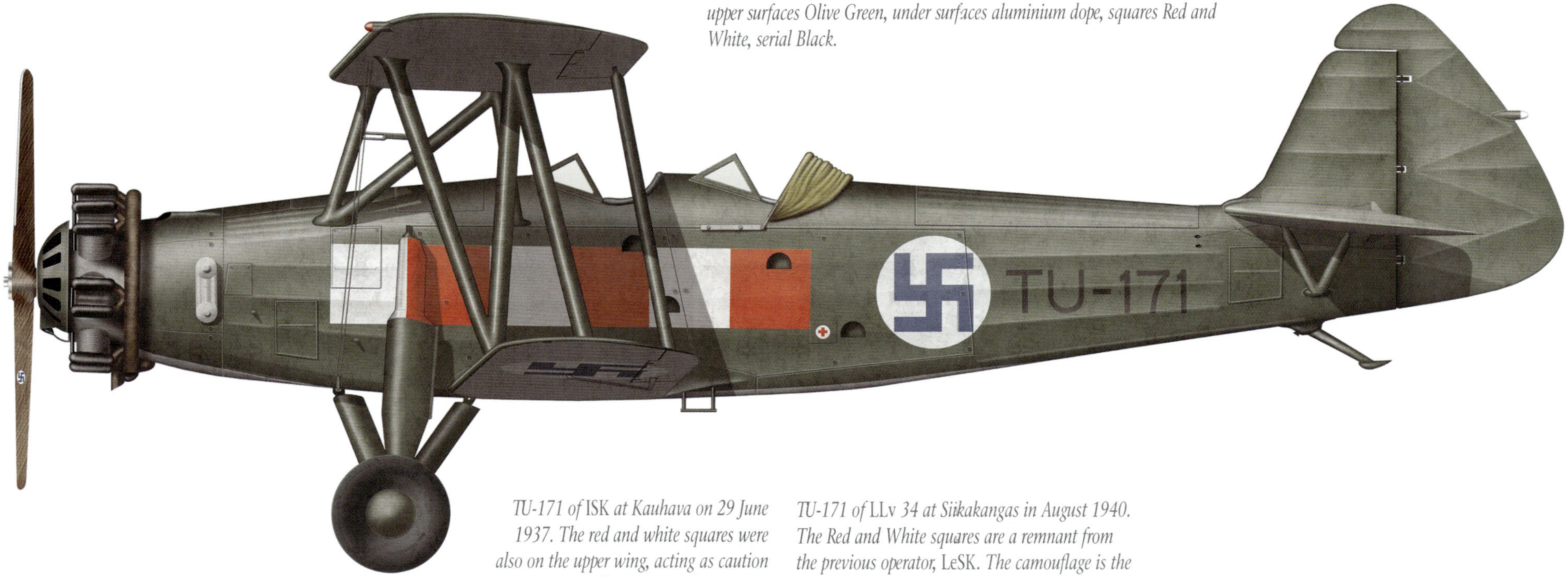

VL Tuisku, TU-171, Ilmasotakoulu, Kauhava, June 1937. Camouflage colours: upper surfaces Olive Green, under surfaces aluminium dope, squares Red and White, serial Black.

TU-171 of LLv 34 at Siikakangas in August 1940. The Red and White squares are a remnant from the previous operator, LeSK. The camouflage is the standard solid Olive Green and aluminium dope.

TU-171 of ISK at Kauhava on 29 June 1937. The red and white squares were also on the upper wing, acting as caution markings. (Finnish Air Force)

VL Tuisku, TU-163, Ilmavoimien Esikunta, Helsinki Malmi, September 1937. Camouflage colours: upper surfaces Olive Green, under surfaces aluminium dope, inscription White, serial Black.

TU-163 was delivered to IlmavE at Helsinki on 27 September 1937. It bears the inscription Pilven Veikko in honour of the university students who donated the funds. (Finnish Air Force)

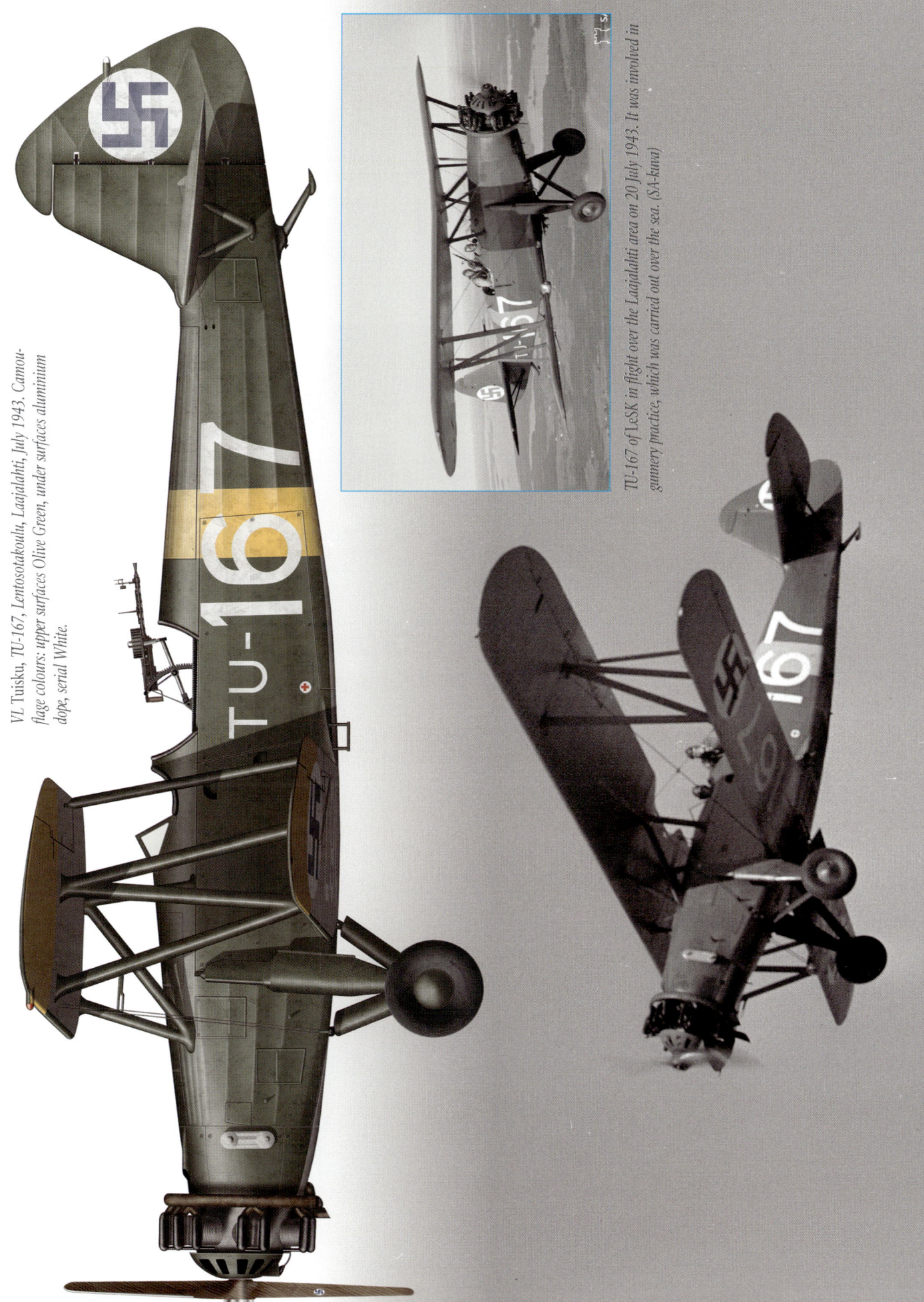

VL Tuisku, TU-167, Lentosotakoulu, Laajalahti, July 1943. Camouflage colours: upper surfaces Olive Green, under surfaces aluminium dope, serial White.

TU-167 of LeSK in flight over the Laajalahti area on 20 July 1943. It was involved in gunnery practice, which was carried out over the sea. (SA-kuva)

Aero A-32GR

Purchase

The Czech company Aero Tovarna Letadel designed in 1923 the A-11 reconnaissance and training biplane, of which 440 examples were built. The Aero A-32GR was a re-engined version from 1927, being powered by a 420 hp Gnome Rhone Jupiter IV radial engines. 116 aircraft with various engines were built.

The Finnish Air Force had eight A-11 aircraft and as a replacement sixteen A-32GR planes were bought in 1928. All but one arrived in Finland between August and October 1929. Due to engine trials the last one came two years later. The aircraft were serialled AEj-49–AEj-64.

Aero A-11 serialled AE-47 of MLE in front of the Utti hangar, having arrived on 12 October 1927. Eight aircraft were bought and they served for almost twelve years. All wore the aluminium dope finish through their whole career. (Finnish Air Force)

Aero A-32 was a re-engined A-11. Here is AEj-53 of MLE in December 1931. The factory finish was overall aluminium dope. The transfer of A-32s to ISK began in late 1937. Most had by the time of the Winter War the Tuisku style camouflage of Olive Green upper and aluminium dope lower surfaces. (Finnish Air Force)

Employment

On arrival most of the aircraft were placed as reconnaissance machines to Land Escadre (*MLE*) at Utti, four planes were with Detached Land Squadron (*ErMLL*) at Suur-Merijoki and one was tested at Santahamina.

When the air stations were formed on 30 June 1933 the Aeros were at Air Station 1 (*LAs 1*) at Utti and *LAs 5* at Suur-Merijoki. Two aircraft eventually went to Aviation School (*IlmK*) at Kauhava.

The regiments were established on 1 January 1938, most of the Aeros serving with the Air Fighting School (*ISK*) in observer and gunner training while individual aircraft were employed as target tugs.

AEj-61 of MLE *on a visit to Suur-Merijoki in July 1930. It was not uncommon that the registration of Aeros was decorated; here the Black serial was turned into tiger stripes. (Finnish Air Force Museum)*

Aero AEj-52 of ISK at Kauhava on 12 August 1936. It has an experimental camouflage of two greys and green with light grey lower sides. This camouflage lasted until 8 October 1936, when the aircraft suffered a bad landing. (Finnish Air Force)

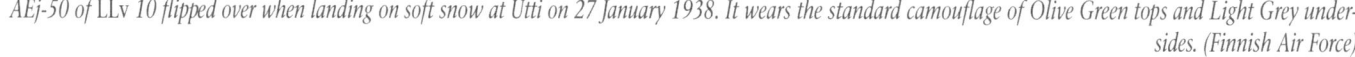

Aero AEj-64 of LLv 10 during war games at Pälkjärvi on 12 March 1937. The participants were marked by broad White bands on upper surfaces and Black bands on lower surfaces. (Finnish Air Force)

AEj-50 of LLv 10 flipped over when landing on soft snow at Utti on 27 January 1938. It wears the standard camouflage of Olive Green tops and Light Grey undersides. (Finnish Air Force)

By the Winter War, commencing on 30 November 1939, ten Aeros remained in service with the Air Fighting School and supplementary units. At the beginning of the Continuation War on 25 June 1941 the Air Fighting School had eight Aeros, but after the beginning of 1942 only three remained airworthy. One was stored the following year and the career of the type ended in the crash of AEj-50 on 30 June 1944.

Colours

All Aeros were originally painted in overall aluminium dope.

After the outbreak of the Second World War an order was given on 12 October 1939 to camouflage trainers similarly to the *Tuisku*. This was done in a major repair and applied to all types.

Seven Aeros, AEj- 50, 55, 57, 58, 60, 63 and 64 were painted Olive Green on top and Light Grey on undersides in late 1939 and early 1940. In 1941 the remaining three Aeros, AEj-51, 52 and 59, received these colours. This scheme remained as long as the Aeros flew.

S/n	Delivered	Struck off charge	Remarks	Hours
AEj-49	24 Aug 1929	7 Mar 1934	W/o Perkjärvi 22 Jun 1933	314.30
AEj-50	24 Aug 1929	15 Jul 1944	W/o Mustasaari 30 Jun 1944	1397
AEj-51	24 Aug 1929	11 Oct 1941	W/o Kauhava 19 Aug 1941	1748.15
AEj-52	24 Aug 1929	11 Feb 1944	W/o Kauhava 19 Jan 1944	1730.30
AEj-53	24 Aug 1929	3 Mar 1938	W/o Kauhava 25 Jan 1938	1184.50
AEj-54	24 Aug 1929	16 Aug 1935	W/o Kauhava 16 Jul 1937	652.35
AEj-55	18 Oct 1929	2 Nov 1940	Into storage 13 Jan 1940	1261.15
AEj-56	23 Oct 1929	29 Apr 1936	W/o Utti 23 Dec 1935	697.35
AEj-57	23 Oct 1929	11 Oct 1941	W/o Kauhava 19 Aug 1941	1540.05
AEj-58	23 Oct 1929	28 Feb 1943	Into storage 4 Feb 1943	1572.35
AEj-59	23 Oct 1929	9 Aug 1944	Into storage 27 Jul 1944	1378.10
AEj-60	23 Oct 1929	5 Oct 1943	Into storage 28 Jun 1941	1018.25
AEj-61	23 Oct 1929	16 Oct 1934	W/o Valkeala 24 Aug 1934	387.40
AEj-62	30 Sep 1929	21 Jan 1936	W/o Tienhaara 18 Dec 1935	706.35
AEj-63	30 Sep 1929	31 Dec 1941	W/o Kauhava 18 Nov 1941	1319.40
AEj-64	11 Sep 1931	11 Oct 1941	W/o Haapamäki 8 Sep 1941	1557.40

AEj-52 of LeSK at Kauhava in November 1941. The colours are the regular Olive Green and Light Grey, with the Yellow Eastern Front markings. On Aeros the undersides of both wing tips were Yellow. The rudder came from a multicoloured plane. (Authors collection)

Aero A-32, AEj-63, Maalentoeskaaderi, Utti, August 1930. Aluminium dope overall, serial Blue with Black shadow.

AEj-63 of MLE in front of Utti hangar on August 1930. It was assigned to the reconnaissance squadron of the unit. It also shows one style of several coloured serial numbers. (Author's collection)

Aero A-32, AEj-52, Ilmailukoulu, Kauhava, August 1936. Camouflage colours: upper surfaces Olive Green and Medium Grey, under surfaces Light Grey, serial Black.

Aero AEj-52 of ISK at Kauhava on 12 August 1936. It has an experimental camouflage of two greys and a Green with Light Grey lower sides. This camouflage lasted until 8 October 1936 bad landing. (Finnish Air Force)

Aero A-32, AEj-60, Lentolaivue 10, Utti, December 1937. Camouflage colours: upper surfaces Olive Green, Medium Grey and Light Grey, under surfaces Light Grey, serial White.

Aero AEj-60 of LLv 10 on its back after a landing at Utti on 21 December 1937. It has another experimental camouflage. (Finnish Air Force)

218

Aero A-32, AEj-59, Lentosotakoulu, Kauhava, March 1942. Camouflage colours: upper surfaces Olive Green, under surfaces Light Grey standard Eastern Front markings Yellow, serial Black.

AEj-59 of LeSK at Kauhava in March 1942. All colours and markings are by the book. In a regular inspection on 27 July 1944, this machine was classified as ready for write-off. (Author's collection)

Blackburn Ripon IIF

Purchase

The State Aircraft Factory bought one Blackburn Ripon pattern aircraft, serialled RI-121, and obtained a licence for local manufacture. The air force also placed an order for fifteen planes, with factory designation VL R.29 Ripon, in two batches, with serials RI-129–RI-143 inclusive.

The pattern aircraft RI-121 arrived in Finland on 20 September 1929, equipped with a 480 hp Jupiter VI radial. For the Finnish-built machines the engine choice was still ahead. The first four aircraft of the I series (RI-129–RI-132) were all fitted with different engines and built by January 1931.

RI-129 was the first licence-built Ripon in Finland. It is seen here at Santahamina on 30 December 1932, belonging to MeLAs. It has the contemporary factory standard serial on the overall aluminium dope finish. (Finnish Air Force)

RI-159 was the last licence-built Ripon in Finland. It is seen here at Santahamina on 16 October 1934, ten days before hand-over to LAs 4. The overall finish is also aluminium dope, but the serial conforms with the 20 March 1934 regulation. (Finnish Air Force)

The remaining three aircraft of the I series (RI-133–RI-135) and the 8-aircraft II series (RI-136–RI-143) were fitted with the Panther radial. They were constructed between August 1931 and February 1932. Inspired by the British Ripon development the Baffin, the power plant for the 10-aircraft III series (RI-150–RI-159) was a 610 hp Pegasus II radial and these were built between March and September 1934. However, the last one, RI-159, was powered by a 650 hp Hispano Suiza Nbr V-12 engine.

The Finnish Air Force had a total of twenty-six Ripons in its inventory.

RI-137 of LeSK at Kauhava in December 1940. The provisional camouflage of Olive Green on aluminium dope dates back one year, to the Winter War with LLv 16. (Author's collection)

Employment

During the 1930s the Ripons served with every maritime unit. In the Winter War mobilization in October 1939, only two squadrons were equipped with the nearly obsolete Ripon, *LLv* 16 and 36.

The Ripons left *Lentolaivue* 16 in May 1940, when the remaining three aircraft were handed over to a training unit, *T-LLv* 17. Later during the Continuation War a few aircraft served with *LLv* 6, 10, 12 and 15 in a number of roles, mainly in maritime surveillance duties.

The dedicated trainer use of the Ripon ended on 10 July 1941, when the last plane was transferred to another squadron. Thereafter solitary Ripons were used as hacks with a few squadrons.

The last flight of the Ripons in Finland was made by RI-156 which, totally worn out, was flown to the air depot on 16 February 1945 and was not struck off until 1950.

Colours

When new in their career the Ripons were painted overall with aluminium dope. This surface finish was still on the planes when the Winter War started. During the mobilization in October and November 1939 the bright finish of several planes was roughly smudged with a brush to obtain a better camouflage effect against aerial reconnaissance.

Though the standard Finnish Air Force camouflage of Olive Green upper sides and aluminium dope lower sides had existed since January 1934, the first Ripon to receive this colour was RI-121 in January 1940, but the lower surfaces being light grey for the wood and fabric skinning.

It was followed next month by RI-156 and RI-157. Eventually four more, chronologically RI-152, 129, 151 and 153, were so painted in 1940 and two more, RI-134 and 140 in summer 1941. Trainer Ripons did not serve long enough to receive the Warpaint of Olive Green and Black.

RI-121 of LeLv 6 at Helsinki Malmi in March 1943. This plane was used in flying courier missions for its unit. The trainer colours of Olive Green and Light Grey camouflage were applied at the factory on 15 July 1942. (Author's collection)

RI-134 of T-LeLv 35 at Vesivehmaa in August 1942. It served as a target-tug, the winch locating under the national insignia. The solid Olive Green and Light Grey camouflage was applied at the factory in July 1941. (Olli Riekki)

RI-156 of HLeLv 26 was used also to deliver mail to the islands in Lake Ladoga. Here it is on a stop at Lahdenpohja on 8 April 1944. The Warpaint with DN-colour undersides was applied exactly one year earlier. (SA-kuva)

Serials

By the Winter War the Ripons wore the standard serials as specified on 20 March 1934, the height of the serial was 2½ times of the arm of the fuselage swastika. The style complied with the Finnish standard SFS Z.I.1.

S/n	C/n	Delivered	Struck off charge	Remarks	Hours
RI-134	I/6	28 May 1940	4 Aug 1944	Last flight 3 Jul 1944	
RI-137	II/9	28 May 1940		To overhaul 10 Jul 1941	
RI-138	II/10	24 May 1940		To LLv 44 1 Jul 1941	
RI-151	III/17	21 Jun 1941		To LLv 15 7 Jul 1941	
RI-153	III/19	21 Jun 1941		To LLv 15 6 Jul 1941	

This table contains only those Ripons which flew in the training role during the given period.

RI-156 of HLeLv 26 in April 1944. Left: outside Valamo in Lake Ladoga on skis and right: a few days later on wheels at Kilpasilta. The plane wears a typical Warpaint of Olive Green and Black with DN-colour underside. (Kauko Tuomikoski)

Blackburn Ripon IIF. RI-138, Lentolaivue 16, October 1939. Camouflage colours: aluminium dope overall with brushed over Olive Green, serial Black.

RI-138 of LLv 16 at Värtsilä in late October 1939. During the Winter War mobilization the bright aluminium dope surfaces were heavily brushed over with Olive Green, to obtain a workable camouflage. (Author's collection)

Blackburn Ripon IIF, RI-137, Lentosotakoulu, Kauhava, July 1941. Camouflage colours: aluminium dope overall with brushed over Olive Green areas, standard Eastern Front markings Yellow, serial Black.

RI-137 of LeSK at Kauhava on 10 July 1941. The camouflage effect of brushing Olive Green over the aluminium dope finish was done in the Winter War by the current user, LLv 16. (Finnish Air Force)

Blackburn Ripon IIF, RI-121, Lentolaivue 6, Helsinki Malmi, May 1943. Camouflage colours: upper surfaces Olive Green, under surfaces Light Grey, standard Eastern Front markings Yellow, serial Black.

RI-121 of LeLv 6 at Helsinki Malmi in May 1943. It was flown on 25 August 1943 to the air depot for storage as totally worn out. (Author's collection)

Blackburn Ripon IIF. RI-138, Koelaivue, Tampere, September 1942. Camouflage colours: upper surfaces Warpaint of Olive Green and Black, under surfaces DN-colour, standard Eastern Front markings Yellow, serial Black.

RI-138 was the hack of KoeLv, which was based at Tampere next to the State Aircraft Factory. The unit received this machine on 30 July 1942 and flew it until a crash on 10 May 1943 (Finnish Air Force Museum)

Fokker C.V

Purchase

The C.V was a Dutch army co-operation aircraft design from 1924. It was built by Fokker and under its licence in several versions, for a total of 956 examples. Finland bought one C.VE for evaluation purposes and it arrived on 20 September 1927 with serial FO-39.

In March 1934 thirteen C.VEs were bought from Fokker. These arrived between March and June 1935 and were serialled FO-65–FO-77.

C.VE serialled FO-39 of IlmK at Kauhava on 14 October 1933. This is an early model, which was bought for evaluation purposes. The overall finish was aluminium dope with natural metal nose throughout the whole service career. (Finnish Air Force)

The first four Fokker C.Vs arrived at Suur-Merijioki on 20 March 1935 to equip LLv 12. Here is the brand new FO-65 on skis. It has typical factory details, Black cowling and Fokker logo on the fin. (Finnish Air Force)

At the beginning of the Winter War Sweden donated three C.VEs, which arrived on 23 December 1939 and received serials FO-19, 23 and 80.

Two Norwegian C.VDs escaped the German occupation and were interned on 9 June 1940. They received serials FO-65 and 66 again, the previous ones being lost earlier. The Finnish Air Force had thus nineteen C.Vs in its inventory.

Employment

The Fokkers served in their intended role with most army co-op squadrons throughout the wars, main users being *LLv* 14 and 16.

The dedicated trainer role of the C.V began on 19 February 1940 with the arrival of two aircraft to *T-LLv* 17. By June 1940 six more had arrived. The fifteen month long intermediary peace was used in intense bomber and reconnaissance aircrew training. Due to flying accidents the C.V trainers were exhausted by mid-September 1941.

Solitary planes served then as squadron hacks or in target and target-towing duties until the end of the hostilities.

Colours

The first plane, FO-39, wore an overall aluminium dope finish throughout its career.

The main batch bought from Holland carried the Fokker factory camouflage of Dark Brown upper and aluminium dope lower surfaces. In subsequent major overhauls or repairs the factory replaced the colours with the regulation Olive Green and Light Grey. The first to be painted thus was FO-68 in October 1937, being followed by eight others by 1941.

The Warpaint of Olive Green and Black, which was introduced on 30 September 1940, was first applied to FO-71 on 22 August 1941. This year saw further application of the Warpaint to FO-65, 66, 70 and 69. FO-68 and 77 followed in early 1942.

The Light Blue-Grey DN-colour was introduced to the underside of warplanes on 7 May 1942. FO-19 and 70 was the first ones to receive this paint on 16 September 1942. Next year it was applied to to FO-76, 65, 69, 77 and 80. FO-66 was the last to get the full Warpaint with DN-colour on 17 March 1944.

Serials

The Fokker C.Vs wore the standard serials as specified on 20 March 1934, the height of the serial was 2½ times of the arm of the fuselage swastika. The style complied with the Finnish standard SFS Z.I.1. The serial was originally White, but in factory visits it was changed to Black.

The Fokkers of LLv 12 in a parade held at Suur-Meri-joki on 3 August 1935. The closest machines are FO-66, 67, 68 and 70. All wear the standard Fokker factory finish of Dark Brown tops and aluminium dope bottoms. (Finnish Air Force)

Fokker FO-72 of LLv 12 at Suur-Merijoki on 15 June 1936. It is very clean with typical factory colours and markings. By the Winter War this unit had swapped its planes for Fokker C.Xs. (Finnish Air Force)

FO-69 of LLv 12 at Suur-Merijoki on 16 July 1937. It has the Fokker applied camouflage of Dark Brown upper and aluminium dope lower surfaces. This is a later model C.VE, distinguished by one large landing gear support. (Finnish Air Force)

FO-76 was the liaison aircraft of bomber squadron LLv 42. It is seen here on a stop at Helsinki Malmi on 14 June 1941, just three days before the Continuation War mobilization. (Author's collection)

FO-76 was the liaison aircraft of bomber squadron LLv 42. It is seen here on a stop to Helsinki Malmi on 14 June 1941, just three days before the Continuation War mobilization. (Author's collection)

FO-19 of LeSK taxied into a ditch and nosed over at Kauhava on 13 June 1941. The regulation camouflage of Olive Green and Light Grey was brand new. (Finnish Air Force)

FO-71 of T-LLv 17 waiting for delivery at Kauhava on 29 July 1941. The first T-LLv 17 was disbanded a week earlier and the new operator was LLv 16. The standard Olive Green upper and Light Grey lower surfaces were painted four weeks earlier. (Finnish Air Force)

FO-71 of LLv 10 ready for take-off at Tiiksjärvi shortly before being shot down on 21 September 1941. The Warpaint with Light Grey undersides was done at the factory just one month before. (Author's collection)

FO-80 was another hack of LeLv 42, seen here at Luonetjärvi shortly before its nose over on 20 July 1942. (Author's collection)

FO-77 of LeLv 14 at Tiiksjärvi shortly before it bellied in on 13 August 1942. The warpaint with Light Grey undersides was painted at the factory on 1 May 1942. (Author's collection)

C.VD serialled FO-65 of LeLv 14 in transit at Luonet-järvi on 1 July 1943. The Warpaint with DN-colour underside was done at the factory on 30 March 1943. (Author's collection)

S/n	C/n	Delivered	Struck off charge	Remarks	Hours
FO-19	23	19 Mar 1941		Damaged 17 Jul 1941	
FO-23	39	19 Mar 1941	7 Nov 1941	W/o Pori 15 Sep 1941	
FO-39	5033	13 Sep 1933	18 Aug 1938	W/o Alajärvi 28 Jul 1938	
FO-66		29 May 1940	20 Aug 1940	W/o Karvia 12 Jul 1940	
FO-68		26 Mar 1940		To *LLv* 10 31 Aug 1941	
FO-69		26 Mar 1940		Damaged 28 Jun 1941	
FO-70		20 Jun 1941		Damaged 27 Jun 1941	
FO-71		3 Jul 1940		To *LLv* 16 Aug 1940	
FO-72		9 May 1940	8 May 1941	W/o Lappajärvi 1 Apr 1941	
FO-74		19 Feb 1940	12 Apr 1940	W/o Kyynärjärvi 27 Feb 1940	
FO-76		15 Apr 1940		Damaged 29 Apr 1940	
FO-77		19 Feb 1940		To overhaul 26 May 1940	
FO-80		22 Mar 1940		Damaged 30 May 1940	

This table lists only those Fokker C.Vs which flew in the training role in the given period.

FO-80 of T-LeLv 35 at Vesivehmaa in August 1943. It was used as a target-tow for the squadron's gunnery practices. The Warpaint with DN-colour underside was done at the factory in 1 June 1943. (Author's collection)

Fokker C.VE. FO-77, Lentolaivue 12. Suur-Merijoki, August 1938. Camouflage colours: upper surfaces Dark Brown, under surfaces aluminium dope, cowling Black, serial White.

FO-77 of LLv 12 being re-fuelled during a visit to Parola on 3 August 1938. It carried the original Dutch factory colours, witnessed by the Fokker logo on the fin. (Finnish Air Force)

Fokker C.VE, FO-68, Lentolaivue 14, Räisälä, February 1940. Camouflage colours: upper surfaces Olive Green and White, under surfaces Light Grey, serial White.

FO-68 of LLv 14 taxiing at Räisälä in February 1940. This plane was painted in December 1939 in the Finnish Olive Green tops, which was not ideal on the snow. The unit applied provisional winter camouflage with a chalk and glue mixture to its aircraft. (Author's collection)

Fokker C.VE, FO-68, Lentolaivue 14, Tiiksjärvi, March 1942. Camouflage colours: upper surfaces Olive Green, under surfaces Light Grey, standard Eastern Front markings Yellow, serial White.

FO-68 arrived with LLv 14 at Tiiksjärvi in February 1942. Quite exceptionally this plane has the Yellow nose, which was specified only for single-engine fighters on 1 September 1941. (Toivo Vuorinen)

Fokker C.VE, FO-68, Lentolaivue 14, Tiiksjärvi, March 1942.

Fokker C.VE, FO-77, Lentolaivue 14, Tiiksjärvi, August 1942. Camouflage colours: upper surfaces Olive Green and Black, under surfaces DN-colour, standard Eastern Front markings Yellow, serial Black.

FO-77 of LLv 14 after a bad landing in the dark at Tiiksjärvi on 13 August 1942. The Warpaint of Olive Green and Black tops with Light Grey bottoms was applied at the factory on 1 May 1942. (Tapani Lampinäki)

Fokker C.VE, FO-69, Pommituslentolaivue 6, Turku, May 1944. Camouflage colours: upper surfaces Olive Green and Black, under surfaces DN-colour, standard Eastern Front markings Yellow, serial Black.

FO-69 of PLeLv 6 at Turku on 17 May 1944. It is prepared for a training sortie for the anti-aircaft defences of light cruiser Väinämöinen based at Turku naval station. (SA-kuva)

Fokker C.VE, FO-80, Lentolaivue 14, Tiiksjärvi, June 1943. Camouflage colours: upper surfaces Warpaint of Olive Green and Black, under surfaces DN-colour, standard Eastern Front markings Yellow, serial Black.

FO-80 of LeLv 14 ready for take-off at Tiiksjärvi on 6 June 1943. The stay with this unit lasted only six weeks as the plane was transferred to the training role with T-LeLv 35 on 16 July 1943. (SA-kuva)

Fokker C.VE, FO-80, Lentolaivue 14, Tiiksjärvi, June 1943.

Koolhoven F.K.52

Purchase

The F.K.52 was a Dutch light bomber and reconnaissance biplane design from 1937. Only six aircraft were manufactured before Germany invaded the Netherlands. Swedish Count Carl-Gustaf von Rosen bought prototypes 3 and 4 and donated them to Finland. The planes were flown to Finland on 18 January 1940 and given serials KO-129 and 130.

Employment

After overhaul by the State Aircraft Factory, both Koolhovens were delivered to *LLv* 36 by 18 February 1940. Their service continued with *LLv* 6 in the Continuation War, flying reconnaissance and propaganda missions.

KO-130 was missing in action on 16 August 1941 and KO-129 served from 23 September 1941 onwards with *LLv* 16, until handed over to *LeSK* 29 September 1942. The short trainer career ended in a crash on 23 February 1943.

Colours

The Koolhovens had the original finish of aluminium dope, but this was overpainted by the factory in February 1940, replaced by the regulation solid Olive Green tops and Light Grey bottoms.

For use with a front-line squadron Koolhoven KO-129 received Warpaint of Olive Green and Black in mid-Septemeber 1941, keeping these colours until it was destroyed.

F.K.52 serialled KO-130 of LLv 36 inside the hangar at Helsinki Malmi on 5 March 1940. It has all regulation colours and markings, with Olive Green upper and Light Grey lower surfaces. (SA-kuva)

Koolhoven F.K.52, KO-129, Lentosotakoulu, Kauhava, October 1942. Camouflage colours:
upper surfaces Warpaint of Olive Green and Black, under surfaces Light Grey standard Eastern
Front markings Yellow. Serial Black on Olive Green paint and Olive Green on Black paint.

KO-129 of LeSK at Kauhava, after arrival on 27 September 1942 on wheels and on
19 February 1943 on skis. The plane crashed just four days later. (Finnish Air Force)

Koolhoven KO-129 of ErLLv at Helsinki Malmi on 26 June 1940. The camouflage is the specified Olive Green with Light Grey underside. (Finnish Air Force)

Koolhoven KO-129 of LeSK at Kauhava having arrived on 27 September 1942. This Warpaint of Olive Green and Black was applied at the factory on 20 September 1941. The lower surfaces remained Light Grey. (Author's collection)

Liaison Aircraft

- Beechcraft C-17L Traveler
- Cessna C-37 Airmaster
- Desoutter II
- Fairchild 24G

- Junkers A 50 Junior
- VL *Kotka*
- Fieseler Fi 156 K-1 *Storch*
- Polikarpov U-2

Beechcraft C-17L Traveler

The C-17L was a private donation from Denmark, arriving on 10 March 1940. The plane was stored until 18 August 1942. It was then given the serial BC-1 and became a liaison plane for the air force HQ. It was written-off on 27 January 1945 after 230 hours logged.

Beechcraft BC-1 of IlmavE at Helsinki Malmi in April 1943. The air force HQ used this plane for top brass liaison through the rest of the war. (Author's collection)

BC-1 of IlmavE on a stop at Jämijärvi in autumn 1943. The Beechcraft was the luxury transporter of its time. (Author's collection)

Beechcraft C-17L, BC-1, Ilmavoimien Esikunta, Helsinki Malmi, June 1943. Overall finish Cream with red decorations, standard Eastern Front markings in Yellow, serial Black.

BC-1 of IlmavE in front of Rissala's hangar on 25 January 1945. Two days later this plane had a stall landing and was damaged. The Yellow identification markings were removed four months earlier. (Author's collection)

BC-1 of IlmavE next to Utti hangar in early 1944. The ski arrangement was fixed. (Author's collection)

Cessna C-37 Airmaster

A C-37 (OH-VKF in the civil register) was appropriated from Veljekset Karhumäki Oy on 25 October 1939. It was given the serial CE-1 and flown as a hack by various units until 30 May 1941, when it was put into storage having logged 234 hours. It was sold back to Karhumäki (an aircraft manufacturer) on 1 December 1943.

Cessna CE-1 of LLv 6 at Santahamina on 23 May 1940. This unit flew on maritime duties, hence the floats. (Finnish Air Force)

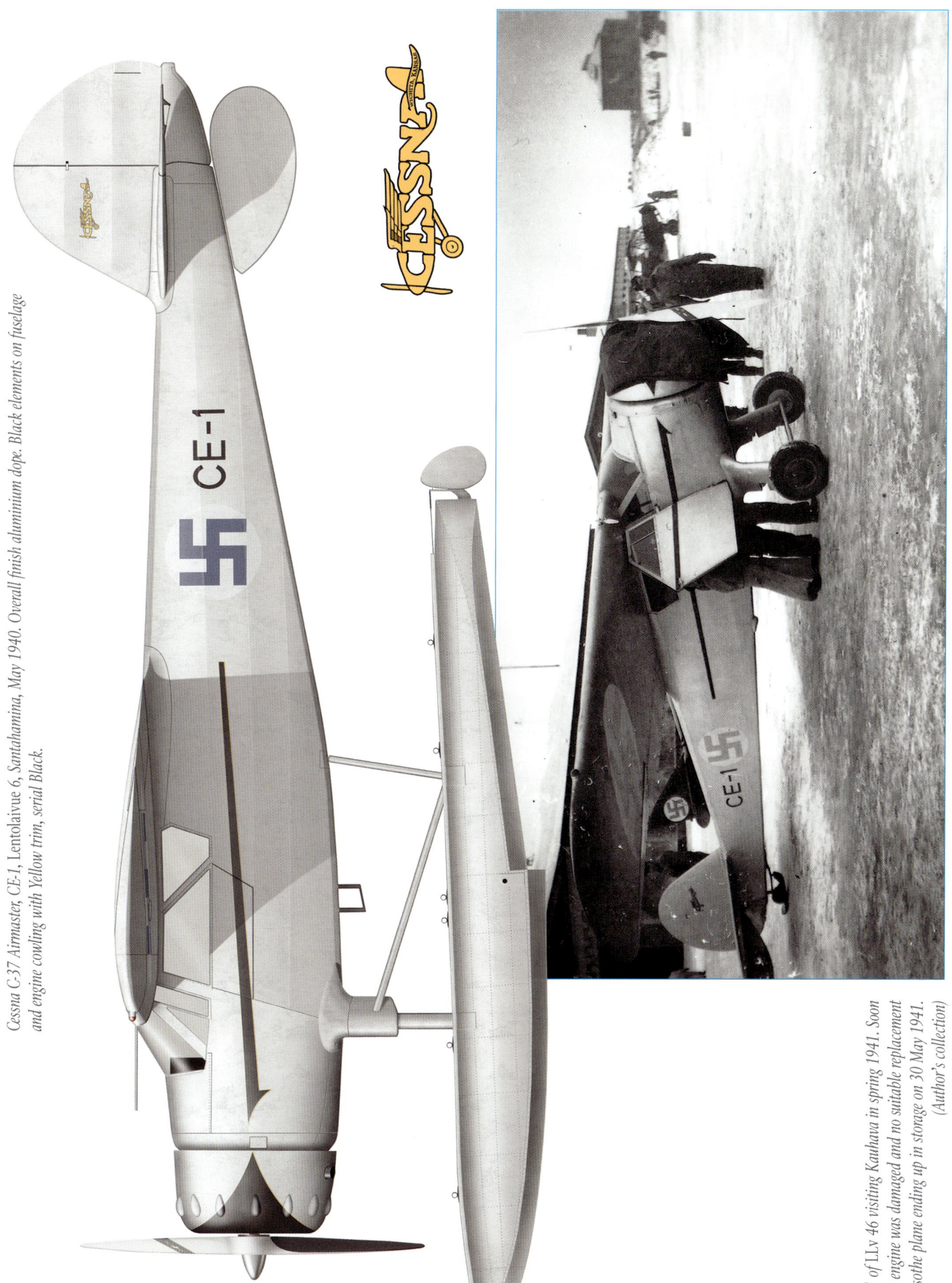

Cessna C-37 Airmaster, CE-1, Lentolaivue 6, Santahamina, May 1940. Overall finish aluminium dope. Black elements on fuselage and engine cowling with Yellow trim, serial Black.

Cessna CE-1 of 1.Lv 46 visiting Kauhava in spring 1941. Soon after this the engine was damaged and no suitable replacement was found, so the plane ending up in storage on 30 May 1941. (Author's collection)

Desoutter II

The Danish Red Cross donated one Desoutter II aircraft, which arrived in Finland on 28 October 1941. With serial DS-1 it flew as a liaison plane of *LeR* 4 squadrons until 14 November 1944, sustaining damage on landing, having logged 475 hours in the air force.

Desoutter DS-1 serving as a hack of LeLv 42 seen in front of a hangar at Luonetjärvi in spring 1943. (Author's collection)

DS-1 of LLv 48 on a visit to Immola in spring 1942. (Author's collection)

DS-1 of PLeLv 44 at Kemi in October 1944, a month after the removal of the Yellow markings. (Author's collection)

Desoutter II, DS-1, Lentolaivue 48, Onttola, April 1942. Overall finish aluminium dope, standard Eastern Front markings Yellow, serial Black.

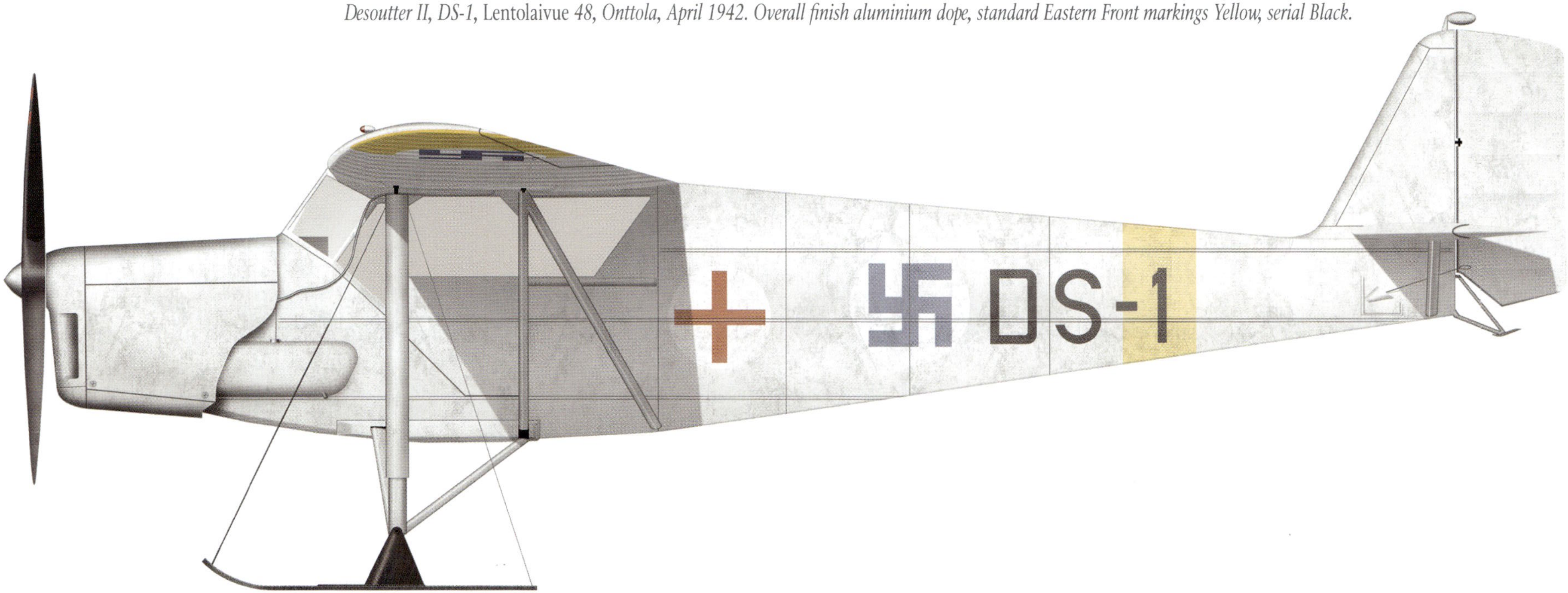

Desoutter DS-1 when with LLv 48, seen here on skis at Onttola in spring 1942. The Red Cross marking behind the cockpit was a carry over of the donor, the Danish Red Cross. (Author's collection)

Fairchild 24G

Fairchild 24G (OH-RIM in civil register) was appropriated on 17 October 1939 and bought on 23 December 1940. As FD-1 it flew on liaison duties with the air force HQ until a crash on 19 April 1941. It logged 221 hours in the air force.

Fairchild coded FD-1 of IlmavE seen on skis in early 1941. The previous owner was Veikko Rimminen, a well known future fighter pilot and ace. (Author's collection)

Fairchild 24G, FD-1, Ilmavoimien Esikunta, Helsinki Malmi, April 1941. Overall Orange-Red. Black elements on fuselage, and engine cowling with Silver trim.

Junkers A 50 Junior

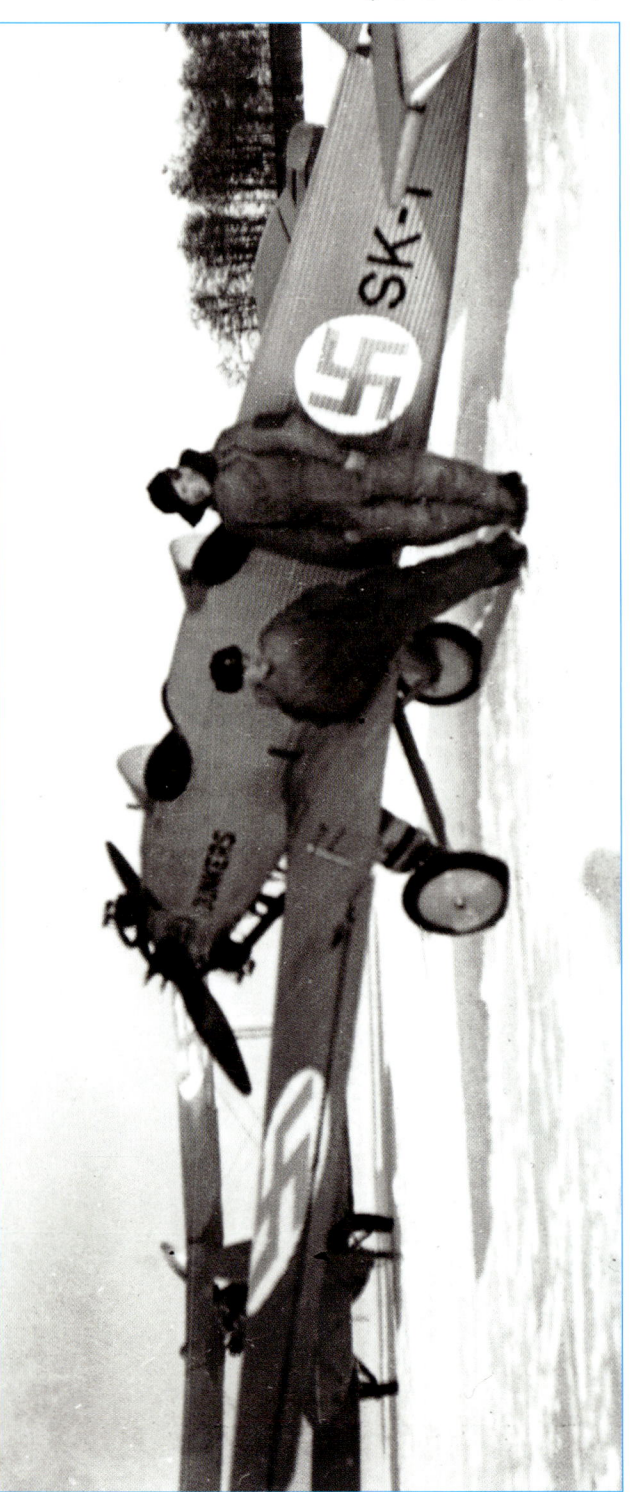

Junkers A 50 Junior, SK-1, Lentolaivue 36, Santahamina, March 1940. Overall finish is dull natural metal.

Junkers A 50 Junior (ex-OH-SKY) of the civil guard was taken over by the air force on 27 April 1936 as SK-1. It was damaged beyond repair with LLv 36 on 12 March 1940, after 204 hours logged. (Author's collection)

JUNKERS

SK-1

VL *Kotka*

The *Kotka* was a State Aircraft Factory (VL) designed multi-purpose aircraft of 1931 vintage, with six planes built. All served on liaison duties and as hacks with various squadrons before the hostilities. During the war four *Kotkas* remained on the same duties. By February 1945 all Kotka flying was over.

S/n	C/n	Delivered	Struck off charge	Remarks	Hours
KA-145	II/1	21 Oct 1939	6 Nov 1941	W/o Pori 27 Sep 1941	
KA-147	II/3	20 Nov 1939	6 Sep 1943	W/o Vesivehmaa 9 Jul 1943	
KA-148	II/4	22 Oct 1939	24 Jan 1945	Into storage 26 Feb 1945	
KA-149	II/5	17 Jul 1941	19 Sep 1942	Into storage 23 Apr 1942	

Aluminium dope finished Kotka KA-145 was the hack of T-LLv 35 and is seen here at Parola in June 1940. (Author's collection)

Olive green camouflaged Kotka KA-149 liaison of LLv 42 in front of Luonet-järvi's hangar in July 1940. (Author's collection)

A mixed camouflaged Kotka KA-148 of T-LeLv 35 at Vesivehmaa in summer 1942. The white fin denoted a target aircraft for Fokker D.XXI gunnery courses. (Author's collection)

255

VL Kotka II, KA-147, Lentolaivue 38, Sortavala, December 1936. Overall finish aluminium dope, serial Black.

Kotka KA-147 of LLv 38 parked at Sortavala air base in Kasinhäntä in December 1936. It went on to serve with six other units until it stalled on landing on 9 July 1943 after 1,518 hours logged. (Finnish Air Force)

VL Kotka II, KA-149, Lentolaivue 42, Luonetjärvi, March 1940. Camouflage colours: unit-applied camouflage of Olive Green and Dark Grey upper surfaces, under surfaces aluminium dope, serial Black with White trim.

Kotka KA-149 of LLv 42 seen at Luonetjärvi on 7 March 1940. It continued in service with five other squadrons until 23 April 1942, when flown into storage as totally worn out after 1,442 hours logged. (SA-kuva)

Fieseler Fi 156 K-1 *Storch*

Finland bought two Fieseler Fi 156 K-1s from Germany, arriving in May 1939. They were given serials ST-112 and 113. The type was evaluated by several units and operated as liaison craft, mostly by *Lentolaivue* 14 and the air force HQ, until 1960.

The Fi 156s were ordered with Dark Green tops and Light Grey bottoms, but there is no photographic proof to support this. Here Storch ST-113 of LLv 6 at Santah. mina on 30 June 1940. It wears the overall factory finish of RLM-Grau 02 or Hellgrau 63. (Finnish Air Force)

ST-112 of IlmavE in Finnish Olive Green and Light Grey camouflage, seen after repairs at Tampere in February 1943. (Author's collection)

Storch ST-113 liaison of LLv 44 visiting Helsinki Malmi in late June 1941. Yellow Eastern Front markings were applied in the normal manner over Grau RLM 02 or Hellgrau RLM 63. (Kyösti Partonen collection)

ST-113 of IlmavE in Finnish colours at Helsinki Malmi on 20 July 1942. Two days later it was handed over to LeLv 14, where it crashed on 26 October 1944, after 1,331 hours logged. (Finnish Air Force)

Fieseler Fi 156 K-1 Storch, ST-113, Lentolaivue 6, Helsinki Malmi, June 1940. Overall finish Grau RLM 02 or Hellgrau RLM 63, serial Black.

Storch ST-113 seen here under evaluation with LLv 6 at Helsinki Malmi in June 1940. (Author's collection)

Fieseler Fi 156 K-1 Storch, ST-112, Ilmavoimien Esikunta, Helsinki Malmi, August 1942. Camouflage colours: upper surfaces Olive Green and under surfaces Light Grey, standard Eastern Front markings Yellow, serial Black.

Storch ST-112 of IlmavE parked on the platform at Helsinki Malmi in August 1942. It flew with this unit until sold for civil use on 31 May 1960 after 2,561 hours logged. All markings are strictly according to the existing regulations. (Author's collections)

261

Polikarpov U-2

Five Polikarpov U-2 biplanes were captured intact and put into flying condition for the air force. The first one made a forced landing on 20 December 1939, but was put into storage. Four more were captured during the Continuation War, refurbished and given serials VU-2, 2*, 3 and 4. Two survived the war and were stored in February 1945 and later scrapped.

This U-2 was the second taken on 5 March 1942 and refurbished by the field air depot, given the serial VU-2. It is seen here inside and outside Immola hangar in March 1943 used as a hack by LeLv 24, until its crash on 31 May 1943 after 489 hours logged. (Both author's collection)

S/n	C/n	Delivered	Struck off charge	Remarks	Hours
VU-2	0575	29 Jun 1942	13 Jul 1943	W/o Ruokolahti 31 May 1943	489.40
VU-2		4 Feb 1944	2 Jan 1950	Last flight 24 Feb 1945	113.10
VU-3		7 Mar 1944	2 Jan 1950	Last flight 17 Feb 1945	230.20
VU-4		8 Apr 1944	13 Jul 1944	W/o Tuulos 23 Jun 1944	68.45

The fourth was captured on 10 October 1943 and it became VU-3. It was the hack of TLeLv 12 until being stored on 23 February 1945 after 230 hours logged. It is seen above at Nurmoila in June 1944 and below at Ve-sivehmaa in September 1944. The plane wears Warpaint, as many front-line hacks did. (Both author's collection)

Polikarpov U-2, VU-4, Tiedustelulentolaivue 12, Nurmoila, April 1944. Warpant camouflage colours: Olive upper surfaces Green and Black, under surfaces DN-colour, standard Eastern Front markings Yellow, subdued insignias, serial Black.

This Polikarpov was the last captured on 3 February 1944. Here VU-4 of TLeLv 12 is seen taking off at Nurmoila in April 1944. It was shot down on 23 June 1944 after 68 hours logged. The plane wears Warpaint and subdued national insignias. (Lauri Kippo)

Polikarpov U-2, VU-2, Hävittäjälentolaivue 28, Hirvas, June 1944. Warpaint camouflage colours: Olive upper surfaces Green and Black, under surfaces DN-colour, standard Eastern Front markings Yellow, subdued insignias, serial Black.

This limousine U-2 was the third captured, on 27 September 1943. It became the hack of HLeLv 28 as the second VU-2 and is seen here at Solomanni in May 1944. The plane was flown for storage to the air depot on 23 February 1945, after 113 hours logged. Typical of front-line squadrons the hacks were also camouflaged in Warpaint. (Author's collection)

Camouflage
& Markings

National and Registration Markings

The Finnish national insignia was born on 2 March 1918, when large blue swastikas were painted on the wings of a Thulin typ D monoplane. These millennia-old good luck emblems of Indian origin meant the same for the donor, the Swedish count Eric von Rosen. Four days later this plane arrived with the army of the white forces in Finland.

Due to the lack of more precise application instructions, many different kinds of insignia appeared on aircraft. The registration markings of the aircraft faced the same phenomenon.

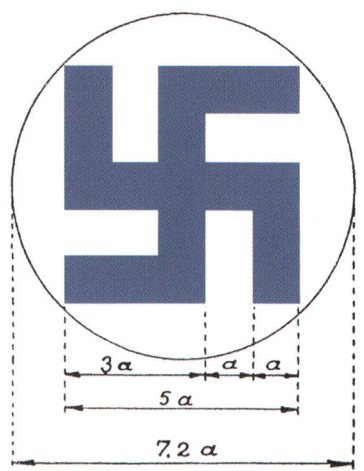

The official proportions of the Finnish national insignia from 20 March 1934 to 1 April 1945.

Tuisku TU-163 of LeSK at Kauhava in late 1941. It displays the large numbers of the serial, which were introduced on 13 May 1939. University students donated the funds for this plane, which received the inscription Pilven Veikko. *(Author's collection)*

On 9 May 1927 a new serial number system was introduced, being still in use. The letter combination, derived from the aircraft's name, informed which particular type was in question. The numbers run for each type consecutively. As time went by various types of lettering appeared on the aircraft.

All variations of the national insignia and registration markings (serial numbers) were put to an end on 20 March 1934, when new permanent marking regulations of the Finnish military aviation SIP IV BA2 were issued. In principle the width of the arm of the swastika defined the size of both the insignia and serial number. The letters and numbers were taken from Finnish standard SFS Z.I.1.

The instructions were clear and simple and they were followed literally, excluding imported aircraft, for the next eleven years.

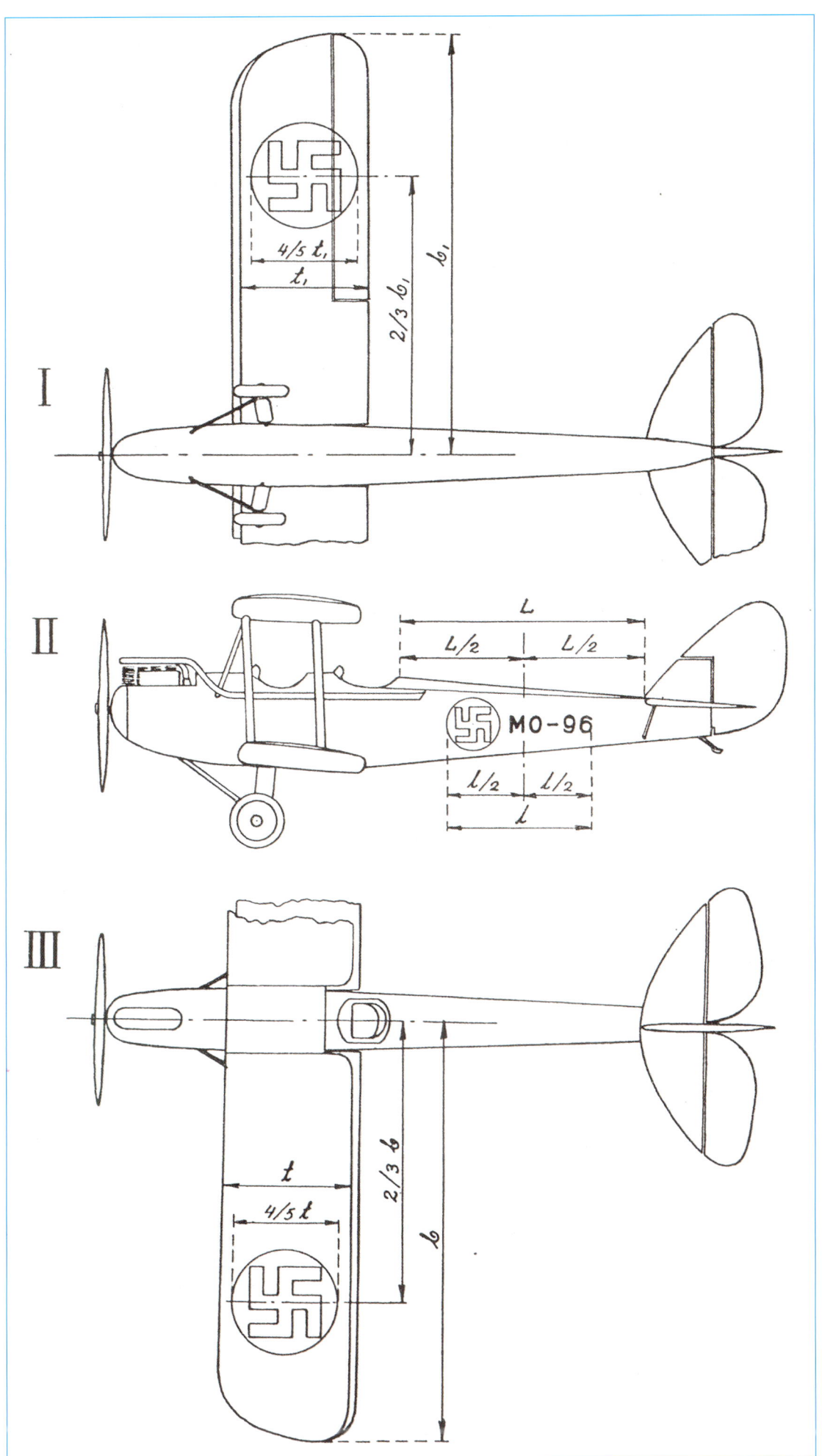

I

II

MO-96

III

Official drawing of the Finnish national insignia and registration markings, concerning their size and location, here on a de Havilland D.H. 60 Moth.

The insignia of large aircraft, especially the Blenheim, were visible over a long distance.

On 23 July 1940 the permanent order was amended slightly, stating that the white circle of the national insignia should not exceed 1 metre. In practise this meant that the wing insignia was 1 metre in size, unless the available space dictated otherwise. Then the 4/5th rule was valid.

On 12 January 1944 an order was issued to subdue the white circle. The factory chose to closest readily-available paint, the DN-väri. The units received from the depot its own mix, which lead to a variety of subdued markings. This regulation concerned only front-line aircraft, but was a carry-over when some aircraft were later relegated to the advanced training role.

S F S	Suomen Standardisoimislautakunta	Finlands Standardiseringskommission	Z. I. 1
Kirjaimet ja numerot kilpiä y.s. varten		**Bokstäver och siffror** för skyltar och dyl.	

abcdefghijklm
nopqrstuvwxy
zåäö
ABCDEFG
HIJKLMNO
PQRSTUV
WXYZÅÄÖ
1234567890
I VIII XV XIII

Ylläkuvatuissa kirjaimissa ja numeroissa on kulmien pyöristys sallittua.
Kirjainten ja numeroiden korkeudet ovat taulukossa, Z. I. 2.

Ovanavbildade bokstäver och siffror kunna även utföras med avrundade hörn.
Bokstävernas och siffrornas höjder se tabell Z I. 2

A page from Finnish standards Z. I. 1 showing the style of the registration letters and numbers (serial number). The height of these was 2½ times the arm of the blue swastika.

After the truce the allied supervision commission, headed by the Russians, laid stress on the Finnish government to remove the swastika insignia as a marking of their prime enemy, the Germans. Therefore the insignia was changed to the Blue and White cockade, effective from 1 April 1945 onwards.

Camouflage

From the 1920s the aircraft in most European air forces were painted in aluminium dope overall. This practise was also followed in Finnish flying units.

The first transfer to camouflage took place in late 1932. The Czech Letov aircraft company, which had delivered a couple of years earlier Š-218 primary trainers to the Finnish Air Force, managed to get an order for a light bomber and army co-op aircraft designated Š-328F. The Ministry of Defence confirmed the order of one prototype, eleven series aircraft and a licence to build 35 planes.

The prototype was to be painted gloss *Kenttävihreä* (Field Green) on upper surfaces and aluminium dope on lower surfaces. The Green also became better known as Olive Green or Standard Green. It was the same hue which was painted on the fuselage tops of licence-built Gloster Gamecock fighters. The colour was an exact match with the British NIVO and French Kaki.

The SK-11 flew for the first time on 19 July 1933 in these colours. Due to various delays and poor flying characteristics, the prototype was not accepted for delivery and the Ministry of Defence cancelled the order the following autumn.

In spring 1933 the State Aircraft Factory received an order for one multi-purpose trainer prototype. It was named Tuisku, with serial TU-149, and made its maiden flight on 10 January 1934. It wore the same colouring as the Letov Š328F, but all later colour references were based on the Tuisku.

Warpaint

During the Winter War and the following busy training season a great many aircraft were damaged and sent to the factory for repairs. Various foreign aircraft were painted with local colours having the closest match. This produced even more hues on the aircraft surfaces.

During summer 1940 new camouflage schemes were tested in both theory and practise by drawings and painting real aircraft. In addition to the standard Green a Dark Grey and Black colour were tried. A directive was issued on 30 September 1940 stating simply that all warplanes are to be painted Olive Green and Black in a 3/2 ratio on top and sides and Silver Grey on lower sides. The Silver Grey meant aluminium dope, but for mainly wooden and/or fabric covered warplanes it was Light Grey.

Pyry PY-2 was the hack of fighter squadron LeLv 28, seen here at Viitana on 25 July 1942. It wears the full Warpaint of Olive Green and Black with Light Grey undersides. The rudder is white from earlier service with supplementary squadron T-LeLv 35. (Author's collection)

Thus a standard camouflage pattern for the Finnish Air Force came into existence. It was named *SOTAMAALAUS*, which simply translates as Warpaint. The aircraft were painted in this scheme at the factory during a repair or major overhaul. The field air depots servicing the front-line aircraft could paint the whole plane when necessary.

The aluminium dope underside was considered too exposing in some front-line units. The comparison became possible when German Dornier Do 17 Z bombers arrived in Finland in early 1942. These aircraft had *RLM* colour *Hellblau* 65 (Light Blue-Grey) undersides.

After short testing the air force headquarters declared that from 7 May 1942 onwards, when painting the undersides of the wings and fuselage of all warplanes, the Light Blue-Grey DN-colour was to be used. This also concerned the VL *Pyry* and Gloster Gauntlet, which were never used operationlly in Finland. Also the *Smolik*, *Viima* and *Steglitz* are known to have carried the DN-colour, when used by a front-line unit. The Warpaint existed until 29 September 1947, when the Black colour was removed during the next re-painting.

Eastern Front Markings

In the middle of May 1941 Germany informed the Finnish military leaders of a forthcoming attack on the Soviet Union, due within a month. With this knowledge a mobilization took place in Finland commencing 17 June 1941.

Next day the Finnish Air Force headquarters informed all flying and anti-aircraft units of new friendly markings for all Finnish aircraft. They were a Yellow 50 cm wide band around the rear fuselage and wing tip underside(s) at the length of 1/6th of the span.

The colour to be applied was a rich yellow colour used by the aircraft factory in small amounts for aircraft fuel piping. This was exactly the same colour as German RLM 04, or British or American Insignia Yellow.

As several hundred aircraft were to be painted in a couple of days, and most of them in the units, neither the air depots nor the paint manufacturer possessed such amounts of the specified colour. A brighter Lemon Yellow replacement colour was applied to the vast majority of aircraft in the squadrons.

From 1 September 1941 onwards the noses of single-engine fighters were to be painted yellow. Depending on the availability the colour could be either of the yellows. Out of advanced trainers the Gloster Gauntlets systematically had the yellow noses and other types at random. It could also be a carry-over from earlier fighter duties.

When the Soviet main offensive commenced on 9 June 1944, just four days later orders were issued for instant removal of the Yellow paint on the upper half of the nose. This process was done by end June 1944, in the units or by the air depots or at the factory.

The Continuation War ended in a truce on 4 September 1944 and all remaining Yellow Eastern Front markings were to be deleted within ten days.

Trainer Colours

At the Air Fighting School a need arose to recognize orange coloured *Smoliks* in flight and on 26 November 1937 one plane (SM-138) had the serial painted as high as could fit on the fuselage, which led to the transfer of the swastika to the rudder. The numbers were also painted in Black under the lower wings.

This served the purpose and on 13 January 1939 all *Smolik* primary trainers were to be marked thus, except the numbers below the wings were in White instead of Black.

On 13 May 1939 the order was extended to cover also *Tuisku* and *Viima* trainers, with *Stieglitzes* added later on. The amendment was to be carried out at a major repair or overhaul.

During the Winter War mobilization the order of painting aircraft was given on 12 October 1939. All new and repaired aircraft were ordered to be camouflaged the way *Tuisku* was, meaning Olive Green upper sides and aluminium dope undersides.

This paintwork was factory applied and in all 15 *Smoliks* and 7 *Viimas* received this type of colour scheme, until on 6 June 1940 new directives were issued.

According to these the *Tuiskus* remained as they were and *Smolik* trainers reverted back to the full Orange exterior. But *Viima* and *Stieglitz* trainers received Olive Green fuselages and Orange wings and tail. Spare parts were allowed to be used in the colour they were.

The amendments were done during a major repair or overhaul and the practise was continued until the end of the careers of these types, the last ones in 1962.

Abbreviations

	Type of aircraft	Total number	Number in training	Main units	Training period	Lost in training	Into storage
BASIC TRAINERS	Letov S218A *Smolik*	39	39	*IlmK, ISK, LeSK*	Mar 1930 – Sep 1944	13	13
	VL *Sääski* I and II	33	33	*IlmK, ISK, LeSK*	Jun 1928 – Sep 1942	24	5
	de Havilland D.H. 60X Moth	23	23	*IlmK, ISK, LeSK*	Feb 1929 – Sep 1944	21	2
	de Havilland D.H. 82 Tiger Moth	1	1	24, 26	Aug 1943 – May 1944	1	
	VL *Viima* I and II	23	23	*IlmK, ISK, LeSK*	Feb 1936 – Sep 1944	4	
	Focke Wulf Fw 44 J *Stieglitz*	35	35	*ISK, LeSK*	May 1940 – Sep 1944	5	
ADVANCED TRAINERS	Gloster Gamecock II	17	9	29, 35, *ISK, LeSK*	Sep 1938 – Jul 1944	9	
	ASJA *Jaktfalk* II	3	3	29, *ISK, LeSK*	Dec 1939 – Sep 1944	2	1
	Bristol Bulldog II and IV	19	16	17, 34, 35, *ISK, LeSK*	Dec 1939 – Feb 1944	14	2
	Gloster Gauntlet II	24	24	17, 34, 35, *ISK, LeSK*	Mar 1940 – Sep 1944	10	13
	Polikarpov I-15bis	5	5	34, 35, *ISK, LeSK*	Feb 1940 – Sep 1944	0	5
	VL *Pyry*	41	41	17, 34, 35, *ISK, LeSK*	Dec 1939 – Sep 1944	16	
	Fokker D.XXI	97	32	35	Apr 1941 – Sep 1944	6	24
	Fokker C.X	39	2	17	Jul 1941 – Dec 1941	0	
	FIAT G.50	35	18	35	May 1944 – Sep 1944	3	1
TWIN-ENGINE TRAINERS	Avro Anson I	3	3	17, 47	Oct 1936 – Sep 1944	2	
	Ilyushin DB-3M	21	5	46, 48	Mar 1940 – Jun 1942	2	
	Hanriot H.232	3	2	17, 48	Aug 1941 – Sep 1944	0	2
	Bristol Blenheim I	97	13	17, 48	Feb 1942 – Sep 1944	4	
	Airspeed Envoy	1	1	17, 48	Apr 1942 – Jul 1943	1	
	Tupolev SB and USB	24	3	17	Feb 1943 – Sep 1944	0	
GUNNERY TRAINERS	VL *Tuisku*	31	31	17, *ISK, LeSK*	Sep 1935 – Sep 1944	11	
	Aero A-32	16	16	17, *IlmK, ISK, LeSK*	Aug 1929 – Jul 1944	12	4
	Blackburn Ripon IIF	29	5	35, *ISK, LeSK*	May 1940 – Jul 1944	0	
	Fokker C.VE and C.VD	19	13	17, 35, *ISK, LeSK*	Sep 1933 – Sep 1941	4	
	Koolhoven F.K. 52	2	1	*LeSK*	Sep 1942 – Feb 1943	1	
LIAISON	Beechcraft C-17L Traverel	1		*IlmavE*	Sep 1942 – Sep 1944		
	Cessna C-37 Airmaster	1		*IlmavE*	Oct 1939 – Dec 1943		
	Desoutter II	1		42, 44, 48	Oct 1941 – Sep 1944		
	Fairchild 24G	1		*IlmavE*	Oct 1939 – Apr 1941	1	
	Junkers A 50 Junior	1		36	Apr 1936 – Mar 1940	1	
	VL *Kotka*	6	4	17, 34, 35	Oct 1939 – Sep 1944	4	
	Fieseler Fi 156 K-1 *Storch*	2	2	6, 12, 14, *IlmavE*	May 1939 – Sep 1944	1	
	Polikarpov U-2	4	4	12, 24, 28	Jun 1942 – Sep 1944	2	

All figures shown here are between January 1930 and September 1944, when the truce with the Soviet Union ended all training.

In the main unit column the number informs the squadron, 17, 25, 29, 34, 35 and 47 being supplementary squadrons

Finnish	Abbreviation	English	Notes
Erillinen	Er	Detached	
Hävittäjälentolaivue	HLeLv	Fighter Squadron	From 14 Jan 1944
Ilmailukoulu	IlmK	Aviation School	To 1 Jan 1938
Ilmasotakoulu	ISK	Air Fighting School	From 1 Jan 1938
Lentoasema	LAs	Air Station	To 1 Jan 1938
Lentolaivue	LL, LLv, LeLv	Squadron	
Lentorykmentti	LentoR, LeR	Aviation Regiment	From 1 Jan 1938
Lentosotakoulu	LeSK	Aviation Fighting School	From 3 Jan 1941
Merilentoeskaaderi	MeLE	Naval Aviation Escadre	To 30 Jun 1933
Merilentoasema	MeLA	Naval Air Station	To 1 Jan 1938
Merilentolaivue	MeLL	Naval Squadron	To 30 Jun 1933
Maalentoeskaaderi	MLE	Land Escadre	To 30 Jun 1933
Maalentolaivue	MLL	Land Squadron	To 30 Jun 1933
Pommituslentolaivue	PLeLv	Bomber Squadron	From 14 Jan 1944
Tiedustelulentolaivue	TLeLv	Reconnaissance Squadron	From 14 Jan 1944
Täydennyslentolaivue	T-LLv, T-LeLv	Supplementary Squadron	From 10 Oct 1939